Formal language aspects of natural computing

Ramanujan Mathematical Society Lectures Notes Series

Vol. 1: Number theory

Vol. 2: The Riemann zeta function and related themes

Vol. 3: Formal language aspects of natural computing

Vol. 4: Commutative algebra and combinatorics

Vol. 5: Convexity in discrete structures

Vol. 6: Number theory and discrete geometry

Vol. 7: Discrete mathematics

Vol. 8: Lectures on operator theory

Vol. 9: Essays on geometric group theory

Vol. 10: Techmüller theory and moduli problems

Ramanujan Mathematical Society

Lecture Notes Series

Volume 3

Formal language aspects of natural computing

Proceedings of a research-level group discussion held at the Indian Institute of Technology, Chennai, November 2005

Volume editors

Kamala Krithivasan
R. Rama

International Press
www.intlpress.com

Ramanujan Mathematical Society
Lecture Notes Series, Volume 3
Formal language aspects of natural computing

Volume Editors:
Kamala Krithivasan
R. Rama

Published in 2010 by International Press, Somerville, Massachusetts, U.S.A. under license from the Ramanujan Mathematical Society.

This work is also published in the Republic of India, exclusively by the Ramanujan Mathematical Society.

ISBN 978-1-57146-188-9

Typeset using the LaTeX system.
Printed in the United States of America.

Preface

Computing with molecules has been a fascinating area of research for long since Nobel laureate Richard Feynman gave his visionary idea in late 1950's. The field of biologically inspired computing has coexisted with main stream computing since the 1930's and the pioneers in this area include McCulloch W. Pitts, R. Rosen, A.M. Turing, J. Von Neumann. These ideas were however first realized only in 1994 when L. Adleman took an instance of a Hamiltonian path problem and solved the problem in a biological laboratory using operations on DNA strands. Since then lot of progress has been made in the area of Natural Computing. DNA computing, membrane computing, peptide computing, computing with ciliates are some of the approaches in Natural Computing. Various theoretical models have also been constructed based on these concepts of Natural Computing. The area is just in an embryonic stage. There is a lot of potential for this area to grow. As Leonard Adleman said "They were built by 3 billion years of evolution and we are just beginning to tap their potential to serve non-biological purposes. Nature has given us an incredible toolbox, and we are starting to explore what we might build." A good amount of work in this area is being carried out in Europe, USA, Canada, etc.

To elaborate further, 'natural computing' is computing inspired by nature. This field of research tries to imitate nature in the way it computes and designs new models and paradigms of computing used by nature. Research in this field includes theoretical and empirical understanding of these paradigms. Thus while evolutionary algorithms use the concepts of mutation, recombination and natural selection in biology, neural networks are inspired by the highly interconnected neural structures in the brain and the nervous system. Another aspect of natural computing is molecular computing which includes DNA computing and peptide computing.

Molecular computing is concerned with the use of biomolecules for the purpose of computing. It has the potential to resolve two well recognized obstacles of conventional computers, miniaturization and massive parallelism. Through molecular computing one descends to the nano-scale computing which solves the miniaturization problem. Since for example, a single drop of solution can contain trillions of DNA molecules, and when an operation is performed on a tube containing DNA molecules, it is performed on every molecule in the tube, massive parallelism is obtained. The well known field of molecular computing, viz DNA computing started with Adleman's paper. A lot of research on this is being done in USA, Israel and Europe. Even though the initial enthusiasm about replacing a silicon computer by this model has come down, it is perceived that DNA computing will be very useful for solving hard problems such as breaking a cryptocode or in DNA to DNA computations. Peptide computing is another model of molecular computing which is still at the conceptual level only.

The idea that molecular system can perform computation is not new and was started with Van Neumann's discussions of self-reproducing automata (Cellular automata) in the 1940's. Molecular computing has opened up new formal models of computations like splicing systems, Watson Crick automata and Sticker systems.

Observing the reactions taking place in a cell, the concept of cellular computing has been developed. It is concerned with the use of membranes, biomolecules and biochemical processes for the purpose of computing. DNA computing can give unbounded parallelism but the concept of membrane computing has been developed to have a distributed and parallel model of computing inspired from the structure and functioning of living cells. At present, it is basically a theoretical model only. Implementation using biomedia or silicon media is being explored.

To popularize this emerging field of natural computing among the Indian research community and attract the young potential researchers to this dynamic field, "A Research level group discussion on Natural Computing" under the sponsorship of CARDMATH (Centre for Advancement of Research in Discrete Mathematics), Department of Science and Technology, India was organized from November 21-27, 2005 at Indian Institute of Technology Madras, Chennai.

This group discussion stressed much on DNA computing, membrane computing, peptide computing and was attended by almost all the leading peer researchers of the Natural Computing area in India which includes Prof. K.G. Subramanian (MCC, Chennai), Dr. Dersanambika (Kerala), Dr. S.N. Krishna (IIT Bombay) Dr. K. Lakshmanan (TIFR, Bombay), Prof. T.S. Chandra (Biotechnology, IITM).

The group discussion had a pack of lectures in the morning session and lab visits and discussion in the afternoon. Some of the existing theoretical results were presented. Students (Ph.D and B.Tech.) from IIT Madras presented some new ideas and were thrown open for discussion.

Research material which highly motivate young researchers were distributed to the participants. As a continuation of our efforts to make this field known, we have attempted to bring out a complied volume in the form of lecture notes. Since the areas that were discussed are not much known to researchers all over India, we have included 11 survey articles, 7 research papers to show further direction of research and three short notes on molecular computing algorithms for specific problems designed by the B.Tech. students motivated by this group discussion.

The volume has two sections followed by a "Students column". The first section contains 11 survey articles covering P systems, DNA computing, network of evolutionary processors, cellular automata, evolutionary algorithms, peptide computing, splicing systems. The second section has 7 research articles in the area of Natural Computing. "Students column" deals with designing algorithms in molecular computing models to solve some NP complete problems on the lines of Adleman's work.

We wish to place on record our sincere thanks to the international academicians who have contributed to the discussion meeting and also to the current volume. Especially we thank Tom Head, Gheorghe Păun, Erzsébet Csuhaj-Varjú, György Vaszil, Helmut Jürgensen, Pierluigi Frisco, Wim Hordijk for their contribution to this volume.

Our thanks to CARDMATH DST, India for sponsoring the programme. Our sincere thanks to the Indian Institute of Technology Madras for having permitted us to host this programme. Our thanks also are due to the Heads of the Department of Mathematics and Computer Science and Engineering for their constant support and encouragement. We also thank our research scholars H. Ramesh, L. Jeganathan, Rita Brata Sen Gupta, Masilamani for their cooperation in organizing this programme. We acknowledge the kind services of Mr. E. Boopal for the compilation work.

March 2006

Kamala Krithivasan
R. Rama

Table of contents

Formal Language Aspects of Natural Computing– 2006, pp. 1–5

One Mathematician's Tour from Biology into Computing and Back to Life

Tom Head

Abstract

A recently retired mathematician reviews his fifty year sight-seeing trip from biology through computing and back. Perhaps researchers entering the new fields that have arisen as composites of the biomolecular and computational sciences may find tributary references that are worth exploring.

In spite of the grandeur of the sweeping research program provided to biology by the theory of evolution, for many of us studying sciences before the mid twentieth century, the biological sciences appeared not to allow sufficiently rigorous and incisive theoretical development. Many of us thought that if one had an excellent memory for detail and a penchant for classification then one might find the biological sciences congenial; but if one wished to come to grips with hard reality then one would be required to study physics, or perhaps chemistry. Biology might eventually be developed as an epiphenomenon of these two real sciences. This naive view ignored the information dimension of genetics. Ultimately an informational volcano erupted at mid century.

In 1953 two papers, each only two pages long, appeared that pulled the biological sciences into center stage during the remainder of the century. These were:

James D. Watson (1928-) and Francis H.C. Crick (1916-2004), *Structure of deoxyribose nucleic acids*, Nature 171(1953) 737–738; and

Stanley L. Miller(1930-), *A production of amino acids under possible primitive Earth conditions*, Science **117**(1953)528–529.

Each of these papers prompted remarkable predictions, or at least startling questions: Life from a test tube - how soon? Not soon: a half century later, no life from a test tube yet. However the first of these papers set off marvelous developments: the deciphering of the genetic code, the development of genetic engineering, and now genomics - with the genomes of many species sequenced and others in progress. The second paper stimulated: the development of cosmochemistry, the search for extraterrestrial life, and the initiation of the newly named science of astrobiology. The biological sciences are now seen to be saturated with information - with that information being processed with astonishing speed and accuracy. Biology, at least at the nano and micro scales, can now be viewed as information represented in and acted upon by an incredibly highly evolved system of biochemistry. Rigorous and incisive results are now obtained with a biology that stands on, not merely the sciences of chemistry and physics, but on the third, relatively new, science of information.

A revolution in linguistic theory was also in process in 1953 sparked by Noam Chomsky (1928-) whose doctoral thesis,

The Logical Structure of Linguistic Theory, U. Pennsylvania, Philadelphia,

was completed in 1955. Chomsky's work on generative grammar and the concurrent concern with the specification of the syntax and compiling of programming languages provided the crucial motivation for the development of Formal Language Theory.

Meanwhile the Russian educated physicist Nicolas Rashevsky (1899-1972) had established at the University of Chicago a school of mathematical biophysics with its associated journal Bulletin of Mathematical Biophysics. In the context of this school, an abstract relational biology was initiated by Robert Rosen (1934-1998), whose extensive and provocative bibliography is currently maintained at the web address:

`http://www.people.vcu.edu/~mikuleck/rosen.htm`.

Rosen's work began with

A relational theory of biological systems, Bull. Math. Biophys. **20**(1959)245–260,

and continued immediately with a sequence of three papers that modeled living systems using mathematical machine theory. The first of these papers was:

Abstract biological systems as sequential machines, Bull. Math. Biophys. **26**(1964) 103–111.

Thus, soon after 1953, Rosen was already modeling the informational biopolymers DNA, RNA, and proteins as strings over alphabets of 4, 4 and 20 symbols, respectively, and modeling fundamental biomolecular processes using abstractions provided by computation theory.

Through the 1960s and 1970s fundamental seminal works by three major figures: John von Neumann (1903-1957), Aristid Lindenmayer (1925-1989), and Michael Conrad (1941-2000), appeared.

Von Neumann's work, demonstrating the logical possibility of constructing self-reproducing machines, was in progress by 1941 but its exposition was delayed by the role he played in both World War II and the Cold War. In this work he developed the theory of cellular automata as a formal platform for treating self-reproduction. The exposition of this work appeared posthumously as:

J. von Neumann, *Theory of Self-reproducing Automata*, U. Illinois Press, Chicago (1966).

Work on theoretical and modeling aspects of cellular automata has continued to the present.

Lindenmayer, a developmental biologist, devised parallel generative grammar systems (L-Systems) capable of modeling the stages of growth of algae, flowering plants, and other organisms. This work provided a basis for generative computer graphics and lead to Lindenmayer's being considered as one of the founding fathers of field of Artificial Life. This work began with:

A. Lindenmayer, *Mathematical models for cellular interaction in development, I, II*, J. Theoret. Biology, **18**(1968)280–315.

The lovely web site having address:

`http://www.biologie.uni-hamburg.de/b-online/e28\_3/lsys.htm`

provides an elementary exposition of L-Systems. Google provides further related web addresses. Research in L-Systems has continued and three major expository volumes are available - one oriented toward developmental biology (1975), one toward mathematics (1980) and one toward graphics and modeling (1990).

Conrad's richly seminal mind produced a sequence of publications concerning biology and computation beginning in 1970 and extending to current posthumously appearing works with joint authors. Often he studied the possibility of computing with biopolymers using interactions of their three dimensional configurations. This work began with:

M. Conrad, *Information processing in molecular systems*. Currents in Modern Biology **5**(1972) 1–14.

An extensive bibliography of Conrad's publications continues to be maintained posthumously at: `http://kli.ac.at/theorylab/AuthPage/C/ConradM.html`

A close examination of the works of the nine researchers whose names appear in the preceding paragraphs is sure to repay the effort required and may suggest new insights to the reader. The thought of these nine scientists provided the background for this mathematician's side-trip from biology into molecular computing and back again to biology. The previous paragraphs supply context to which researchers beginning the study of computation with biomolecules (DNA-computing) may relate their new science. We conclude this article now with a skeleton of more recent references that touch each of the central developments relating computation and biopolymers as seen by the author.

The action of enzymes on biopolymers can, in at least a few cases, be so precise that the action can be modeled meaningfully as a formal operation on strings of symbols over an appropriate alphabet (of four nucleotides, or 20 amino acids). The cut and paste operation of restriction enzymes and a ligase on dsDNA was so modeled in:

T. Head, *Formal language theory and DNA: An analysis of the generative capacity of specific recombinant behaviors*, Bull. Math. Biology **49**(1987)737–759

which provides new analogs of generative grammars called splicing systems. The original goal of splicing theory was to provide an algorithm for deciding whether any specified dsDNA molecule can arise from any given finite list of varieties of dsDNA acted on by a given set of restriction enzymes and a ligase. This goal has been achieved, but further efforts are in progress to make the process practical. Can the splicing formalization of the action of restriction enzymes suggest further algorithmic characterizations of enzymatic actions involving polymers? This was the author's hope. Several authors have shown that, by governing splicing by appropriate formal control mechanisms, a basis for universal computation can be obtained.

The data on which various algorithms act to produce the the solution of a given problem instance (for example, a Directed Hamiltonian Path Problem) can often be encoded into ssDNA molecules having well-chosen sequences. Placed in a solution with a ligase, the ssDNAs may anneal into dsDNAs in such a manner that one or more of the dsDNAs can be isolated from the solution and decoded to provide the answer to the problem. DNA-computing was born when a 7-vertex, 14-edge DHPP instance was solved in a wet lab using standard techniques of biotechnology, as reported in the revolutionary paper:

L. Adleman, *Molecular computation of solutions of combinatorial problems*, Science **226** (1994)1021–1024.

Notice the beauty and suggestiveness of Adleman's idea: Molecules are allowed to follow their nature and self assemble into structures that encode problem solutions. This work has initiated engineering-like studies using self assembly and has contributed to new developments in nanoscale technology.

Computing using self assembly, splicing systems, and other formal mechanisms suggested by molecular biology are among the topics treated at great length in the fundamental reference volume:

G. Paun, G. Rozenberg and A. Salomaa, DNA Computing: *New Computing Paradigms*, Springer-Verlag, Berlin (1998),

which is an exposition of the initial stage of DNA-computing.

During the initial phase of DNA-computing there was a degree of optimism about the possibility that one or more of the following three features of DNA-computing might allow competition with standard (currently silicon based) computing: the truly vast parallelism of the biomolecular operations (trillions of distinct polymers being acted on simultaneously), the extreme compactness of information storage in biopolymers (one bit per cubic nanometer in dsDNA), and the minuscule energy used per chemical-bond-altered. DNA to surpass silicon how soon? Not soon. There has been only one computational problem for which the solution is not immediately clear mentally that has been solved (a 20-variable 14-clause SAT in Adleman's lab) and it could have been solved vastly more efficiently on a standard computer. There may now be no residual optimism for competing with silicon, although there may still be hope for a hybrid formed by linking a biomolecular component into a primarily silicon computer. Have the many millions of dollars, yen and euros spent funding research in this area been wasted? Perhaps not. Three major potentialities have appeared on the horizon: (1) the development of new understandings of naturally occurring complex biomolecular processes, (2) the development of computational schemes that can be carried out inside individual cells or among collections of cells, and (3) the development of applications in nanotechnology.

Laura Landweber (Princeton, Ecology and Evolutionary Biology) and Lila Kari (U. Western Ontario, Computer Science) have done distinguished interdisciplinary research on the mind-boggling reorganization procedure carried out inside certain single celled ciliated eukaryotes that results in a new macronucleus being constructed from the DNA of what was previously one of its micronuclei. The nucleotides in the micronucleus are in quite different order than the nucleotides in the macro nucleus. Consequently naturally occurring molecular mechanisms inside the ciliate must perform these remarkable rearrangements (and of course virtually without any errors). The combined expertise of a biologist and a computer scientist has initiated an extensive literature beginning with:

L.F. Landweber and L. Kari, *The evolution of cellular computing*: Nature's solution to a computational problem, BioSystems **52**(1999) 3–13.

Each of the three books listed below includes material dealing with analyses of this naturally occurring computation by ciliates.

A scheme for molecular computing, called aqueous computing, has been carried out in wet labs at universities in Leiden, Binghamton, and Hokkaido. In this scheme molecules in solution, all having the same sequence, are used as blank tablets on which bits can be written by making modifications at specific locations on the molecules. A computation always begins with a sufficiently vast number (perhaps a nanomole) of the molecules on which writing is to be done. This scheme makes fundamental use of the fact that the entire memory can be replicated in unit time (by pouring half a tube into a second tube) and that memories can be merged in unit time (by pouring into a common tube). Addresses are not used for individual molecules in memory. Individual bit locations on the molecules are addressed and altered in parallel on all molecules present in memory. The first aqueous computation was reported in:

T. Head, G. Rozenberg, R.S. Bladergroen, C.K.D. Breek, P.H.M. Lommerse and H. Spaink, *Computing with DNA by operating on plasmids*, BioSystems **57**(2000)87–93.

As with other DNA-computing schemes, the difficulties and error-proneness of the biochemical operations suggest that only small computations, that can be quickly carried out using standard experiments can be done dependably. Although writing with electromagnetic radiation on molecules or molecular complexes has been suggested for aqueous computing, one encounters the problem of finding a sufficiently large set of wavelengths that can be distinguished with sufficient precision by the molecule chosen to serve as the tablet.

While one may not see how large problems can be solved biochemically, one can ask:

Are there circumstances in which small problems are important, but where standard computers cannot be used? Yes, inside organisms - perhaps even inside individual cells, even in bacteria. We have seen that the natural micro-macro DNA shuffling done by ciliates is being viewed and analyzed as a complicated computational achievement. Can we incorporate into a strain of E. coli a simple decision making program that has no output except on the input of a specific toxin produced by a specific microbe, in which case the program triggers the production of a molecule that attacks the microbe? Since E. coli strains are native to the human intestinal tract, we might use harmless strains of E. coli that we have equipped with such defensive programs to solve medical problems. This is merely the shortest example illustrating the value of carrying out tiny computations inside cells. Such examples can be multiplied endlessly. However, one's enthusiasm for these examples might be tempered by the question: Isn't the incorporation of such small programs into cells apt to resolve into exercises in genetic engineering? Is there added value in conceptualizing these questions as involving biomolecular computation. Perhaps the view from computing will stimulate fresh ideas and new activities.

Recall that topic of computing in cells has two aspects: understanding how naturally occurring cells do their own computing, and using cellular chemistry and communications to do our own computing. Each of the following three books concerns one or both of these aspects and is highly recommended for close examination:

A. Ehrenfeucht, T. Harju, I. Petre, D.M. Prescott and G. Rozenberg, *Computation in Living Cells - Gene Assembly in Ciliates*, Springer-Verlag (2004).

Martyn Amos (Ed.), *Cellular Computing*, Oxford U. Press (2004).

J. Chen, N. Jonoska, G. Rozenberg (Eds.), *Nanotechnology: Science and Computation*, Springer-Verlag, Berlin (2006).

An elaborate computing model, called membrane computing (or P-Systems), was suggested by the activities within and the communications among the membrane bound organelles of eukaryotic cells. This model has been developed by Gheorge Paun and his many research collaborators. The literature on membrane computing is already enormous although Paun initiated it only in 1998, and the first major exposition appeared only in 2000. Readers will want to examine closely the following volume of papers that has just appeared.

Ciobanu, G. Paun, M.J. Perez-Jimenez (Eds.), *Applications of Membrane Computing*, Springer-Verlag, Berlin (2006)

A web page for P-Systems is maintained at:

`http://psystems.disco.unimib.it.`

Tom Head
Professor Emeritus of Mathematical Sciences
Binghamton University, Binghamton,
New York 13902-6000 USA
`email: tom@math.binghamton.edu`

Formal Language Aspects of Natural Computing– 2006, pp. 7–15

An Overview of Computation in Cellular Automata

Wim Hordijk

1 Introduction

Cellular Automata (CA) are mathematical models of decentralized spatially extended systems. They consist of a large number of relatively simple individual units, or "cells", which are connected only locally, without the existence of a central control in the system. Each cell is a simple finite automaton that repeatedly updates its own state, where the new cell state depends on the cell's current state and those of its immediate (local) neighbors. However, despite the limited functionality of each individual cell, and the interactions being restricted to local neighbors only, the system as a whole is capable of producing intricate patterns, and even of performing complicated computations. In that sense, they form an alternative model of computation, one in which information processing is done in a distributed and highly parallel manner. Because of these properties, CA have been used extensively to study complex systems in nature, such as fluid flow in physics or pattern formation in biology, but also to study information processing (computation) in decentralized spatially extended systems (natural or artificial). Here, we will give a brief overview of the different ways in which computations can be done with cellular automata.

The paper is organized as follows. First, the concept of cellular automata is introduced. Then some examples of specific computational tasks that can be performed with CA are presented. Next, different proofs of universal computation in CA are discussed. Then, an example of what is called "emergent computation" in CA is reviewed. Finally, some pointers to more information on cellular automata are provided.

2 Cellular Automata

A *cellular automaton* (CA) consists of a regular grid (lattice) of cells (automata), each of which can be in one of a finite number of states. At discrete time steps, all cells simultaneously update their states depending on their current state and those of their immediate neighbors (i.e., depending on the local neighborhood configuration of each cell). Generally neighborhood can be defined in any manner. Local neighborhood means the current cell and physically immediate neighbors. In two-dimension, it can be five neighbored meaning self, top, down, left and right cells or nine neighbored meaning all cells surrounding it including itself. For this update step, all cells use the same deterministic update rule, which lists the new cell states for each possible local neighborhood configuration. This update process is then repeated ("iterated") for a certain number of time steps.

In the simplest case, the CA lattice is a one-dimensional array of cells (1D CA), where each cell can be either black or white (two possible states), and the local neighborhood of a cell consists of the cell itself, the immediate neighbor to the left, and the immediate neighbor to the right (i.e., a *radius* of 1, or three cells total). So, there are $2^3 = 8$ possible neighborhood configurations for a cell, since each of the three cells in the local neighborhood can be either black or white. Such a 1D, two-state, radius 1 CA is known as an *elementary* CA (ECA). Of course many variations on this basic scheme exist, such as higher dimensional

(2D, 3D, ...) lattices, a larger number of states (> 2), or larger neighborhood sizes (a radius of 2, 3,...). However, the basic principles (updating of cells based on local neighborhood configurations according to a given update rule) remain the same in all cases.

The example below shows one particular ECA update rule. The eight possible local neighborhood configurations are listed in the first row (white is represented by a 0, and black is represented by a 1), and the new cell state for each neighborhood is given in the corresponding position in the second row.

```
neighborhood: 111  110  101  100  011  010  001  000
new state:     0    1    1    0    0    0    1    0
```

For this particular update rule, the three neighborhoods "black, black, white" (110), "black, white, black" (101), and "white, white, black" (001) are mapped into a black cell (1), and the other neighborhoods are mapped into a white cell (0). Of course the 0s and 1s in the second row (the new cell states) could have been assigned differently, and there are $2^8 = 256$ ways of constructing an ECA update rule (since each of the eight neighborhoods can be mapped into a 0 or a 1). Figure 1 shows a simple example of how cells in a CA lattice change their states given the above update rule. For example, the second cell in the lattice has a local neighborhood configuration of "white, black, black" (011), and according to the given rule will be white (0) at the next time step (first arrow). The fifth cell has a local neighborhood configuration of "white, white, black" (001), and thus will be black (1) at the next time step (second arrow). Similarly, all cells are changed to their new state simultaneously according to the same update rule given above.

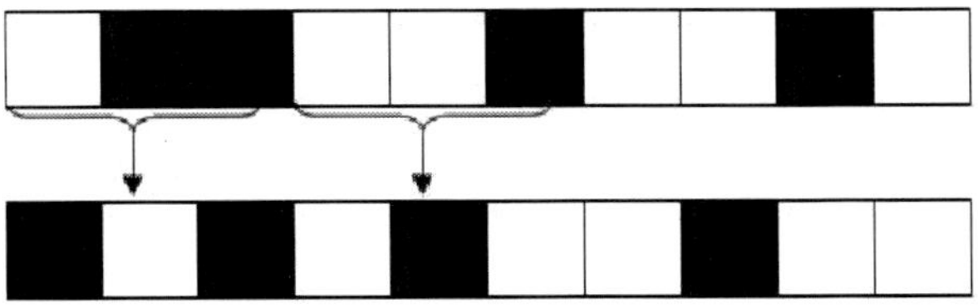

Figure 1: The cells in the CA lattice update according to their own current state and those of their direct neighbors.

In case of a finite lattice, as in the example above, we also need to specify boundary conditions. In this case, *periodic boundary conditions* are used, i.e., the lattice is considered to be circular. Thus, the first and last cells in this array are each other's neighbor, and therefore the first cell has a local neighborhood configuration of "white (last cell), white (cell itself), black (second cell)" (001), and will become black (1) at the next time step according to the given update rule. Alternatively, fixed boundary conditions can be used, where an additional cell is placed on either side of the lattice with a fixed cell state (i.e., these boundary cells are not updated).

Depending on which one of the 256 ECA update rules is used, different behaviors (or dynamics) can be observed when iterating the CA. Figure 2 gives examples of so-called *space-time diagrams* for four different ECA. In these diagrams, the CA lattice is shown horizontally, while time is going down the page. In other words, the first row in each diagram is the CA lattice at the initial time step $t = 0$ (in these examples, the CA lattices are initialized randomly, i.e., for each cell a state, black or white, is chosen at random). The second row is the updated lattice at time step $t = 1$, and so on for each next row. These space-time diagrams show 100 cells across (with periodic boundary conditions), for 100 time steps. Clearly, the four different update rules give rise to very different kinds of dynamics.

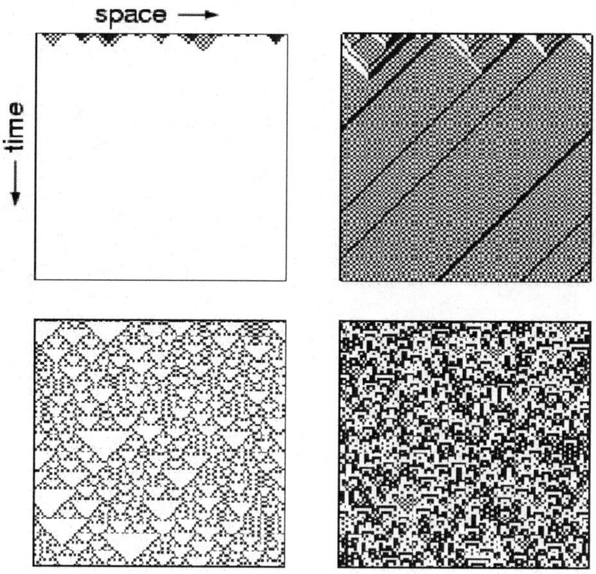

Figure 2: Examples of the dynamics of four different elementary CA.

3 Computation in Cellular Automata

It is possible to use the patterns that form in these cellular automata dynamics to perform computations. Of course not all CA, or all patterns they produce, can be said to perform computations, but sometimes it is possible to construct a specific CA update rule to perform a certain computational task. Some examples are given next.

3.1 Formal language recognition

Smith used CA as formal language recognizers [1]. In particular, one-dimensional, radius 1 "bounded" CA were studied, i.e., CA with two special (fixed state) boundary cells on either end of the lattice. Smith proved that the class of languages that can be recognized by such CA is the class of context-sensitive languages. Some specific examples he included are the context-free language of palindromes (i.e., words that are the same when read from left-to-right and from right-to-left), and the context-sensitive language $\{a^m b^m c^m | m \geq 1\}$. For both cases, he showed the construction of a bounded CA that recognizes one of these languages in linear time. An input word, such as *aabbcc*, is given as the initial configuration to the CA, and it will then decide in linear time whether the input belongs to the language or not. The decision is made by using one particular cell as the "answer cell". If an input word of length n belongs to the language, this particular cell will be in the "accepting" state after n time steps (iterations), otherwise it will be in the "rejecting" state. It is possible for the CA to make this decision in linear time, because it can process the symbols in the input word in parallel.

3.2 Arithmetic

In [2], one-dimensional "filter automata" were used to perform arithmetic. The difference between filter automata (FA) and regular CA is that in FA the cells are updated from left to right, instead of simultaneously. So, when updating a particular cell, the new cell states of neighbors to the left and current cell states of neighbors to the right are considered for the local neighborhood configuration. In this class of automata, often "soliton-like" structures (patterns) can be observed, i.e., propagating periodic structures that can pass through each other without destroying each other, but only shifting each other's phase. Using these types of structures, in [2] a simple "adder" was constructed. Two (binary) numbers are given as input in the initial configuration of the FA, and after a certain number of iterations the lattice will contain the sum of these two numbers. Similar arithmetic operations (such as multiplication) can also be implemented in an FA this way.

3.3 Random number generation

Wolfram gives an example of an ECA that can be used as a pseudo random number generator [3]. For this particular ECA, the state values that are attained by one (arbitrarily chosen) cell in the lattice seem to form a stream of bit values that is indistinguishable from a random stream. In other words, for a large enough lattice, one particular cell in this CA can function as a generator of (pseudo) random bit values. Since the description of an ECA is very simple and concise, this can of course provide many useful practical applications, for example in the area of cryptography, as suggested by Wolfram.

3.4 Image processing

Another, perhaps more practical, task that can be performed with CA is that of image processing. Suppose an image is given as input to a 2D CA, where each cell in the lattice corresponds to a pixel in the image. In other words, for black and white images, if a pixel in the image is white, the corresponding cell in the CA is initialized to white, and similarly for black pixels. For gray-scale images, we can use for example 256 states in the CA, each state corresponding to a particular gray-scale level. In [4], two CA update rules are given for two different image processing tasks: noise filtering and edge detection. It is then shown that using these CA rules, it is indeed possible to remove noise from, or to detect edges in, an given image.

4 Universal Computation in Cellular Automata

The examples above show how CA can be used for very specific computational tasks. However, it has been proved that, in principle, CA can perform any computational task. In other words, they are capable of universal computation. One relatively straightforward way of proving this is by showing that a CA can, in principle, simulate any given Turing machine. This was done for two-dimensional CA in [5]. Obviously, from this it follows that it is possible to construct a 2D CA that simulates a universal Turing machine as well.

However, the actual construction of such a CA, including the correct initial configuration, would be very complicated. Therefore, the next natural question is: "What is the simplest CA that can perform universal computation?" In [6], a one-dimensional, 7-state, radius 1 CA was constructed that can simulate a Turing machine. And finally, after an early conjecture by Wolfram, it was even proved that there is one elementary CA (known as rule 110) that is capable of universal computation [7]. So, even the simplest version of a CA is, in principle, computationally as powerful as any other computing device!

One early, but entirely different, proof of universality in CA was provided using a famous two-dimensional CA known as the "Game of Life". This CA was originally invented by Conway, and is described in [8]. In this CA, there are only two possible states: "dead" and "alive". The update rule is also simple. If a cell is alive, it remains alive if exactly 2 or 3 of its neighbors are alive, otherwise it dies. If a cell is dead, it remains dead unless exactly 3 of its neighbors are alive, in which case it becomes alive. The neighborhood of a cell consists of the 8 cells directly surrounding it (in a 2D lattice). This neighborhood refers to local 9 neighborhood. However, despite this simple rule, the CA shows complicated behaviors and intricate structures, including propagating structures called "gliders". These gliders are small (consisting of 5 alive cells) periodic structures, with a periodicity of 4 time steps, which are displaced one cell horizontally and vertically after one period. So, over time, these structures are propagated through the lattice, and interact with each other upon collision, either annihilating one or more of them, or changing direction, etc. Furthermore, there exist other structures that produce gliders ("glider guns"), destroy them ("eaters"), delay them ("delayers"), or make them change direction. As it turns out, these gliders and other structures can be used in such a way as to simulate any logical function, in particular `AND`, `OR`, and `NOT` gates. Since any logical system containing such functions is universal, it follows that the "Game of Life" is also universal, i.e., capable of universal computation.

5 Emergent Computation in Cellular Automata

In the examples given above of computation in CA (specific or universal), a particular CA update rule and a corresponding initial configuration were constructed, or at least it was shown that such a construction is possible in principle. However, for most computational tasks of interest, it is very difficult to construct a CA update rule or it is very time consuming to set up the correct initial configuration. A very different approach was taken by Mitchell *et al.* in [9, 10, 11], where a genetic algorithm was used to *evolve* a CA rule to perform a specific computational task.

A *genetic algorithm* (GA) is a search algorithm based on the principles of natural evolution (see the introduction to evolutionary computation in these proceedings). It can be used to search through the space of possible CA rules to try to find a CA that can perform a given computational task. The specific task that Mitchell *et al.* originally considered is *density classification*. Consider one-dimensional, two-state CA. For any initial configuration (IC) of black and white cells, we can ask: "Are there more black cells or more white cells in the IC?" Note that for most "computing systems" (including humans), this is an easy question to answer, as the number of black and white cells can easily be counted and compared. However, for a CA this is a difficult task, since each individual cell in the lattice can only check the states of its direct neighbors, and none of the cells can have information of the global state of the CA lattice. So, the question is whether there exists a CA update rule that can decide on the density of black cells (smaller or larger than 0.5) for any IC. In particular, if a given IC contains more black cells (density > 0.5), then the CA should settle down (within a certain maximum number of iterations) to a configuration of all black cells and remain in that state. Otherwise (density < 0.5), it should settle down to all white cells.

A genetic algorithm was used to search through the space of all one-dimensional, two-state, radius 3 CA update rules (of which there are $2^{128} \approx 3.4 \times 10^{38}$, i.e., too many to perform an exhaustive search). The way the different CA were evaluated during the search is as follows. A set of 100 random ICs is generated, and the CA being evaluated is iterated for M time steps on each of these. The lattice size was chosen to be $L = 149$, and M was set to roughly $2L$. The fraction of ICs on which the CA gives the correct answer (i.e., it correctly settles down to all whites or all blacks) is taken as the "fitness" of the CA, a measure of how well it performs the density classification task. Many runs of the genetic algorithm were performed, and in the end the best CA (the one with the highest fitness value) over all these runs was taken as "the solution". Two space-time diagrams of this "best" density classification CA are shown in figure 3. On the left the (random) IC contains more white cells, and on the right it contains more black cells. However, in both cases the CA correctly classifies the density of black cells. (The gray areas are alternating black and white cells, or a "checkerboard" pattern.) Periodic boundary conditions are used.

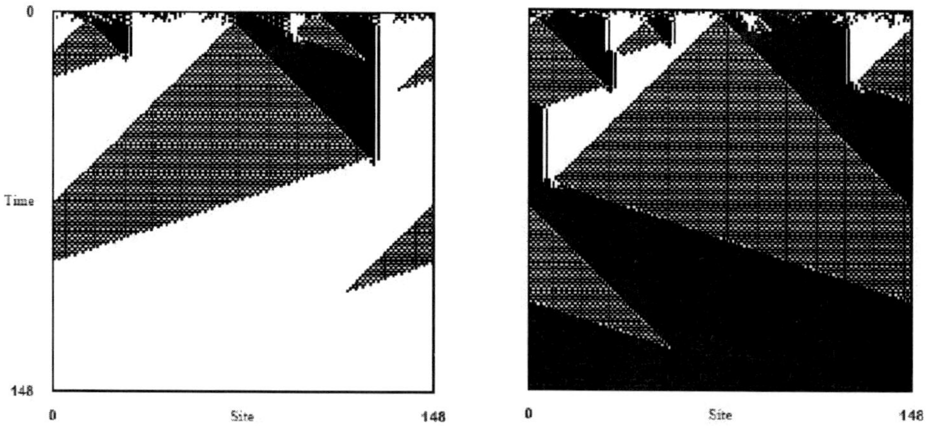

Figure 3: The best CA found by the genetic algorithm for the density classification task.

So how does this CA actually perform the density classification task? As is obvious from the space-time diagrams, the CA dynamics very quickly (after just a few initial time steps) settles down into a collection of very regular patterns, in particular regions of all white, all black, or checkerboard, with sharp boundaries between these regular regions. Furthermore, these boundaries move through the lattice at a certain velocity, and they interact with each other, creating new or destroying existing regular patterns and boundaries. So, it seems that the CA is using these patterns that *emerge* in its space-time dynamics to store, transfer, and process local information (local densities), eventually coming to a global decision about the overall density of the IC.

To appreciate the information processing, or computation, in this CA better, consider the two space-time diagrams shown in figure 4. In both cases, the IC contains one "block" of consecutive white cells, and one "block" of consecutive black cells (since periodic boundary conditions are used, the white region wraps around and thus forms one continuous block). At the boundary between the white (W) region and the black (B) region, a checkerboard (#) pattern is formed, and two boundaries are created (W# and #B), which travel in opposite directions but at the same velocity. At the boundary between the black and white regions, this boundary (BW) persists and remains in the same location over time (i.e., the vertical boundary). In the figure on the left, the original white block in the IC is larger than the black block, and therefore the #B boundary reaches the BW (vertical) boundary well before the W# boundary does. When the #B and BW boundaries meet, the black region and the two boundaries themselves are destroyed, and a new boundary (#W) is created that travels back in the same direction as the original W# boundary, but with a higher velocity. So, eventually the new #W boundary catches up with the existing W# boundary, and upon collision they annihilate each other and the checkerboard region, leaving the entire CA lattice in an all white state. As can been seen in the figure on the right, exactly the opposite happens when the white block in the IC is smaller than the black block. So, in both cases the density of black cells in the IC is classified correctly by the CA (< 0.5 on the left and > 0.5 on the right, with the correct corresponding answers of all white and all black, respectively).

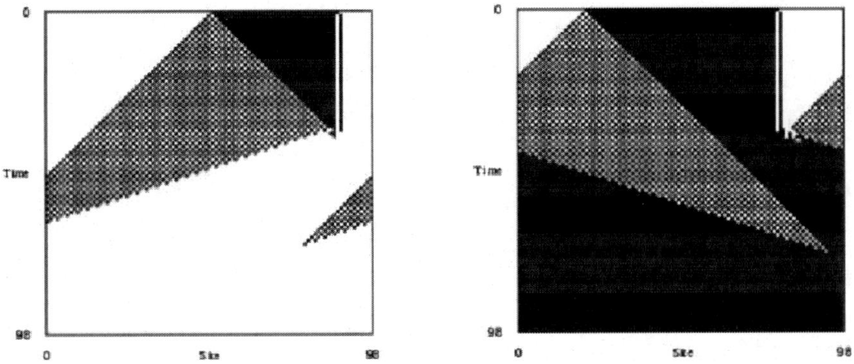

Figure 4: The "strategy" of the evolved CA for solving the density classification task.

Going back to the first pair of space-time diagrams where random ICs were used, it can be seen that this "strategy" of creating checkerboard patterns and using interacting boundaries, as just explained, is used at different time- and length-scales. Local information about densities in small parts of the lattice is stored in patterns of all white, all black, or checkerboard, and the boundaries between these regions transfer this information to other parts of the lattice, where the information is processed through interactions of these boundaries. Decisions about local densities are made this way, and are propagated to other parts of the lattice, until finally a global decision about the overall density of the IC is made. Since the CA makes use of these patterns that (spontaneously) emerge in its dynamical behavior, this type of information processing is generally referred to as *emergent computation* [12, 13, 14].

The above explanation of emergent computation in the evolved CA is a rather informal argument. However, the mechanisms of emergent computation in CA have been studied more thoroughly and have been formalized in a mathematical framework in [15, 16], where a concise model has been developed based on the notion of these propagating and interacting boundaries. In fact, this model (which is a higher-level description of the dynamics of the CA) can be used to make accurate predictions about the CA's behavior and computational performance. Furthermore, it can be used to better understand the actual evolution (by the genetic algorithm) of emergent computation in CA.

6 More Information

Cellular automata were originally introduced by von Neumann after a suggestion by Ulam [17, 18]. They were made popular by a series of articles by Gardner on the "Game of Life" in Scientific American (see [19] for an overview), but were more or less neglected otherwise. In the 80s, with the increase in computing power, research on CA revived again, and Wolfram was one of its pioneers (see [20] for a collection of his papers on CA). Since then CA have been used in many different areas and as models of many different systems. For a more elaborate overview of computation in cellular automata, which is only briefly presented here, see the chapter on this topic in [21]. And finally, many CA resources are available on the internet. A good starting point is `cafaq.com`, and any websearch on the topic will result in numerous additional pages. For more information on evolving cellular automata with genetic algorithms and emergent computation in CA, with links to downloadable electronic versions of publications including [9, 10, 11, 12, 13, 15, 16], please visit `www.santafe.edu/projects/evca`.

References

[1] A. R. Smith: Real-time language recognition by one-dimensional cellular automata. *Journal of Computer and System Sciences* **6** 1972, 233–253.

[2] K. Steiglitz, I. Kamal, and A. Watson: Embedding computation in one-dimensional automata by phase-coding solitons. *IEEE Transactions on Computers* **37** 1988, 138–145.

[3] S. Wolfram. Random sequence generation by cellular automata. *Advances in Applied Mathematics* **7** 1986, 123–169.

[4] A. Popovici and D. Popovici: Cellular automata in image processing. In *Proceedings of the 15th International Symposium on the Mathematical Theory of Networks and Systems*, 2002.

[5] A. R. Smith: Simple computation-universal cellular spaces. *Journal of the Association for Computing Machinery* **18** 1971, 339–353.

[6] K. Lindgren and M. G. Nordahl: Universal computation in simple one-dimensional cellular automata. *Complex Systems* **4** 1990, 299–318.

[7] M. Cook: Universality in elementary cellular automata. *Complex Systems* **15** 2004, 1–40.

[8] E. Berlekamp, J.H. Conway, and R. Guy: *Winning Ways for your Mathematical Plays* **2** Academic Press, 1982.

[9] M. Mitchell, P. T. Hraber, and J. P. Crutchfield: Revisiting the edge of chaos: Evolving cellular automata to perform computations. *Complex Systems* **7** 1993, 89–130.

[10] M. Mitchell, J. P. Crutchfield, and P. T. Hraber: Evolving cellular automata to perform computations: Mechanisms and impediments. *Physica D* **75** 1994, 361–391.

[11] R. Das, M. Mitchell, and J. P. Crutchfield: A genetic algorithm discovers particle-based computation in cellular automata. In *Parallel Problem Solving from Nature - PPSN III*, 1994, 244–253.

[12] J.P. Crutchfield and M. Mitchell: The evolution of emergent computation. *Proceedings of the National Academy of Sciences USA* **92(23)** 10742, 1995.

[13] M. Mitchell, J. P. Crutchfield, and R. Das: Evolving cellular automata with genetic algorithms: A review of recent work. In *Proceedings of the First International Conference on Evolutionary Computation and its Applications*, 1996.

[14] S. Forrest (editor): *Emergent Computation*. North-Holland, 1990. (a special issue of Physica D, vol. 42, nos. 1-3).

[15] W. Hordijk, J. P. Crutchfield, and M. Mitchell: Embedded-particle computation in evolved cellular automata. In *PhysComp96*, 1996, 153–158.

[16] W. Hordijk, J. P. Crutchfield, and M. Mitchell: Mechanisms of emergent computation in cellular automata. In *Parallel Problem Solving from Nature*, 1998, 613–622.

[17] J. von Neumann: *Theory of Self-Reproducing Automata*. University of Illinois Press, 1966. (edited and completed by A. W. Burks).

[18] A. W. Burks (editor): *Essays on Cellular Automata*. University of Illinois Press, 1970.

[19] M. Gardner: *Wheels, Life and other Mathematical Amusements*. W. H. Freeman and Company, 1983.

[20] S. Wolfram: *Cellular Automata and Complexity*. Addison-Wesley, 1994.

[21] T. Gramss, S. Bornholdt, M. Gross, M. Mitchell, and T. Pellizzari: *Nonstandard Computation*. VCH Verlagsgesellschaft, 1998.

Wim Hordijk
Karunya Institute of Technology and Sciences
Deemed University, Karunya Nagar, Coimbatore - 641 114
`email: wim@santafe.edu`

Formal Language Aspects of Natural Computing– 2006, pp. 17–29

On DNA Computing

R. Rama

1 Introduction

We all know, the techniques and concepts of mathematics are applied everywhere. A curious question arises. Is there any field, whose ideas are applied to mathematics? The answer is a humble 'yes'. This marked the first instance of biology being applied to mathematics. Leonard Adleman succeeded in solving an instance of the directed Hamiltonian Path Problem just by manipulating DNA strings. Thus a new area Molecular Computing was born, and later developed into a branch of Natural Computing which studies the computations performed in nature as well as the computations inspired by nature. DNA computing is one of the areas of bio-molecular computing wherein DNAs are the molecules chosen. The idea behind the DNA computing is the simple observation of the following two processes, one biological and one mathematical, are analogous.

1. **Biological Process:** The very complex structure of a living organism is the result of applying some operation (molecular operations such as copying, splicing etc.) to initial information encoded in a DNA sequence.

2. **Mathematical Process:** The result $f(w)$ of applying a computable function to an argument w can be obtained by applying a combination of basic operation to w.

Thus, DNA computing is a fast growing research area connected with the use of either the DNA molecules or the "behaviours" of the DNAs for the implementation of computational processes. It is multidisciplinary by nature, lying in the interface of biology, mathematics, computer science etc. One of the bold goals of this area is to design computers which shall be based on DNAs (rather than on silicon). Although constructing computers from DNA molecules was already suggested by R. Feynman, way back in 1961, and a quite a number of theoretical ideas were published since then, the real beginning of this area was with the publication of the seminal paper "Molecular Computations of solutions to combinatorial problems" by Adleman. The reason is that Adleman was the first one to actually implement a computation in a wet lab and he performed the first "proof-of-principle" experiment for DNA computing.

A naturally occurring DNA strand is a succession of four different building blocks, called bases, arranged in a certain order that depends on the living organism to which the DNA belongs. While in nature, the order in which the bases occur determines the genetic information, these bases can also be chemically synthesized and combined in strands. Instead of "writing" genetic information with the synthetic DNA strands, we "write" the encoding of the input of the problem to be solved and perform the molecular operations (*in vitro*) with the help of molecular biology laboratory and arrive at a DNA sequence which is then decoded as the solution to our problem.

Since the success of Adleman, DNA computing has evolved into an exciting multidisciplinary research area. This area has led already to interesting new computing paradigms, which certainly enriched our understanding of natural computation. As we are aware, understanding of what computation is about is a central goal of (theoretical) computer science. One can be proud enough to feel that DNA computing is already successful in that sense.

2 A brief biological background

For a more detailed description of the notions contained in this section, the reader can refer to the book [6], which gives a nice introduction to molecular biology and genetic engineering with no requirement of background in either biology or bio-chemistry.

2.1 DNA

Proteins are the fundamental agents of life; every human being contains something like 100,000 different proteins. Their properties and their interaction between each other determine the way we are. DNA (Deoxyribo Nucleic Acid) is found in every cellular organism as the storage medium for genetic information, that defines the proteins, the engine of life. DNAs are composed of units called nucleotides, distinguished by the chemical group or bases, attached to them. The four nucleotides that compose a strand of DNA are: Adenine (A), Guanine (G), Cytosine (C) and Thymine (T). Nucleotides are linked together end-to-end to form DNA strands. A short single-stranded polynucleotide chain, usually less than 30 nucleotides long, is called an oligonucleotides.

Every nucleotide consists of three components, a sugar, a phosphate group and a nitrogenous base. The name of the sugar used here is deoxyribose. The sugar has five carbon atoms. Since the nitrogenous base also has carbon atom, the carbons of the sugar are numbered from $1'$ to $5'$. The phosphate group is attached to the $5'$-carbon and the base is attached to the $1'$-carbon. Within the sugar structure, there is a hydroxyl group (OH) attached to the $3'$-carbon. The structure of a nucleotide is illustrated in a very simplified way as in fig. 1.

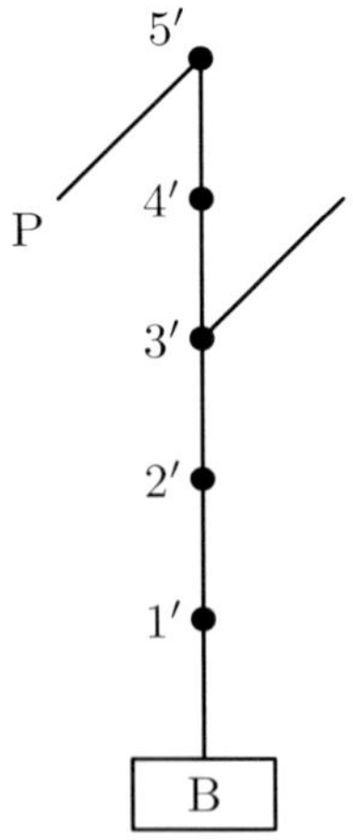

Figure 1:

The $5'$-phosphate group of one nucleotide is found with the $3'$-hydroxyl group of the other forming a phosphodiester bond. A chain of nucleotides are formed as in fig. 2.

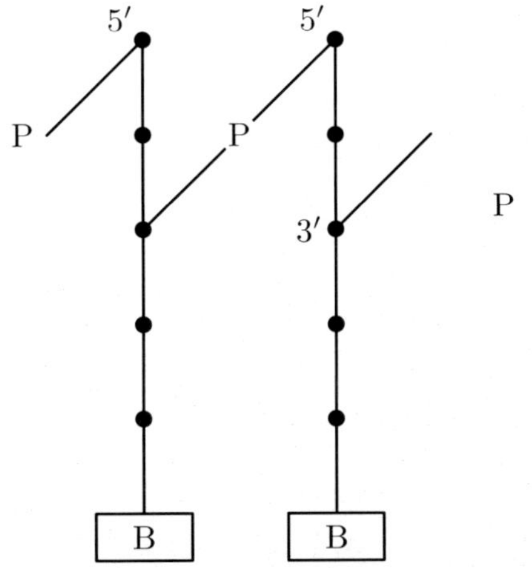

Figure 2:

We represent a chain of nucleotides as $ATGCC\ldots$, by just writing the base of the nucleotides in sequence. This DNA sequence (or single strand DNA) has a polarity. The two distinct ends of a DNA sequence are known under the name of the 5′-end and the 3′-end. If the 5′-end is free in the leftmost nucleotide in a DNA strand and the 3′-end of the rightmost nucleotide is free, then such a DNA strand will have the 5′-3′ direction. Two single-stranded DNA sequences with opposite polarity will join (hydrogen bond) together to form a double-helix (a double stranded DNA) in a process called base-pairing or annealing. This bond follows a property of complementarity. Adenine bonds with Thymine and vice versa and Cytosine bonds with Guanine and vice versa. This is known as Watson-Crick complementarity. The Watson-Crick complement of A is T and vice-versa, the complement of G is C and vice-versa. The complement of a DNA strand is got by replacing a base with its Watson-Crick complement. For example, the complement of (say) $ATTGC$, denoted by $\overline{ATTGC}$, is $TAACG$. The double helix structure of the DNA was discovered by James. D. Watson and Francis H.C. Crick, who were awarded the Nobel prize in 1953. The length of the ssDNA (single strand DNA) is the number of nucleotides present in the strand. The length of the dsDNA (double stranded DNA) is counted in the number of base pairs.

$$ATGGGC \quad \text{6 bps (base pairs)} \tag{1}$$

$$TACCCG \tag{2}$$

$$ATTTTTT \quad \text{7 mers (monomers).} \tag{3}$$

2.2 Operations on DNA

We now describe some of the operations that are available on DNA strands. The purpose of this section is to give an idea of the possible implementation of the algorithm that appears in DNA computing. Some of the simple operations that can be performed on DNA sequences are accomplished by a number of commercially available enzymes that execute a few basic tasks. Enzymes are proteins that catalyze chemical reactions taking place in living cells.

2.2.1 Synthesis

Oligonucleotides may be synthesized by a machine of the size of a microwave oven. The synthesizer is supplied with the four nucleotide bases in solution, which are combined according to a sequence entered by the user.

A desired molecule is built up nucleotide by nucleotide on a support particle (may be a glass surface) in sequential coupling steps. The first nucleotide say A, is bound to a glass support. A solution containing C is poured in and A reacts with C to form AC. This goes on till we achieve the required strand.

2.2.2 Mixing (Union)

Pour the contents of two test tubes into a third one to achieve the union. Mixing can be performed by rehydrating the tube contents (if not already in solution) and then combining the fluids together into a new tube, by pouring.

2.2.3 Annealing (Hybridization) and Melting (Denaturation)

The hydrogen bonding between two complementary sequences is weaker than the one that links nucleotides of the same sequence. It is possible to pair (bond) two ssDNA of opposite directions by cooling the solution containing the DNAs. It is also possible to separate (to debond) a dsDNA into two ssDNAs by heating the solution containing the DNAs to a required level.

2.2.4 Amplifying (Copying)

Make multiple copies of DNA strands by using the "Polymerase Chain Reaction" (PCR). PCR is a machine with which one can make multiple copies of DNAs.

PCR is an *in vitro* method that relies on DNA polymerase (an enzyme) to quickly amplify (make copies of) specific DNA sequences in a solution. PCR involves a repetative series of temperature cycles, with each cycle comprising three stages. Repeating the cycle n times will yield $2n$ copies of a DNA strand. PCR was devised in 1985 by Karry Mullis and Mullis was awarded the Nobel prize in 1985. As Paun et. al., put it, PCR is a very efficient molecular xerox machine.

2.2.5 Lengthening DNA

A class of enzymes called polymerase is able to add nucleotides to an existing DNA molecule. A single strand that has a portion of dsDNA subsequence can be polymerized (extended). Given a short primer oligonucleotide p (the strand to be extended), the enzyme polymerase extends p (always in $5' \to 3'$ direction) if and only if p is bound to a longer template oligonucleotide.

$$\begin{array}{rcl} \text{template} & \text{polymerase} & \\ 5'\,\mathrm{A\,T\,A\,G\,A\,G\,T}\,3' & \longrightarrow & 5'\,\mathrm{A\,T\,A\,G\,A\,G\,T} \\ \mathrm{T\,C\,A}\,5' & & \mathrm{T\,A\,T\,C\,T\,C\,A} \\ \text{primer} & & \end{array}$$

2.2.6 Separation

Gel electrophoresis is an important technique for sorting DNA strands by size. Electrophoresis is the movement of charged molecules in an electric field. Since DNA molecules carry negative charge, when placed in an electrical field, they tend to migrate towards the positive pole. The rate of migration of a molecule in an aqueous solution depends on its shape and electrical charge. Since DNA molecules have the same charge per unit length, they all migrate at the same speed in an aqueous solution. If electrophoresis is carried out in a gel (usually agarose, polyacrylamide or a combination of the two), the migration rate of a molecule is also affected by its size (migration rate of a strand is inversely proportional to the logarithm of its molecular weight). This is due to the fact that the gel is a dense network of pores through which the molecules must travel. Smaller molecules therefore migrate faster through the gel, thus sorting them according to size.

2.2.7 Cutting

There are enzymes which destroy the internal phosphodester bonds in the DNA molecule. They are quite specialised as to what they cut, where they cut, and how they cut.

For example,

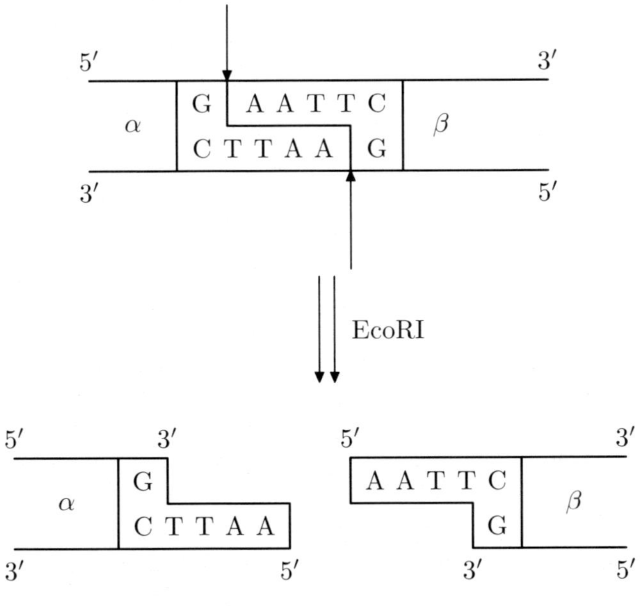

Figure 3:

By cutting a double stranded DNA with the enzyme, we see two double stranded DNAs with sticky ends. With some other enzymes, we can cut the DNAs blunt.

2.2.8 Ligation

DNA molecules may be linked together through a process called ligation which is mediated by enzymes called ligases. The sticky ends develop a hydrogen bonding of complementary base with the help of the enzymes called ligase. For example, the DNAs with the sticky ends can be ligated with the enzyme ligase and we get as in fig. 4

Thus ligation is just pasting DNA strands with compatible sticky ends using ligase.

2.2.9 Detect

Given a test-tube T, the operation 'detect' checks whether T contains at least one DNA strand and say "yes" or "no".

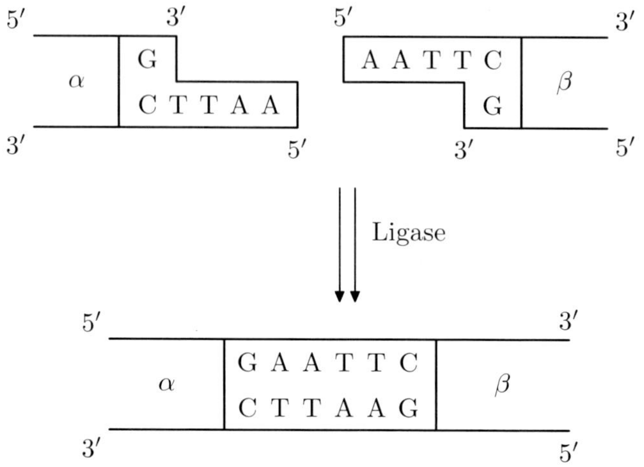

Figure 4:

3 Adleman's experiment

Adleman's experiment described in [1] suggests the feasibility of carrying out computations at the molecular level. The practical possibilities of encoding information in a DNA sequence and of performing simple bio-operations were used by Adleman to solve a seven-node instance of the directed Hamiltonian Path Problem* for the following graph.

Problems can be ranked in difficulty on the basis of the time taken by the best known algorithm (to solve the problem) that will take to execute on a single computer. Algorithms whose running time is bounded by a polynomial (exponential) in terms of the size of the input describing the problem, are in the "polynomial time" class P problems (respectively called exponential time class EXP). A problem is called intractable if it is so hard that no polynomial time algorithm can possibly solve it. The class NP "Non-deterministic polynomial time" class contains the problems for which no polynomial time algorithm (for solving the algorithm) is known but that can be solved in polynomial time by using a non-deterministic computer (a computer that has the ability to pursue an unbounded number of independent computational searches in parallel). An NP-complete problems are the ones that has the property that every other problem in NP can be reduced to it in polynomial time. Thus, in a sense, NP-complete problems are the "hardest" problems in NP.

*A directed graph G with designated vertices V_{in} and V_{out} is said to have a Hamiltonian Path if and only if there exists a path that begins in V_{in} end at V_{out}, and enters every other vertex exactly once

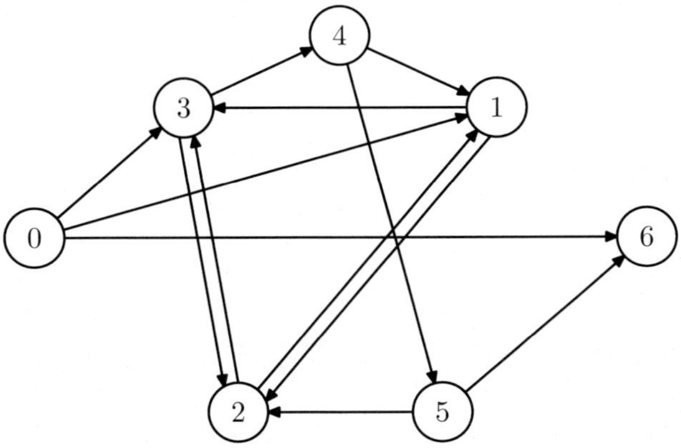

Figure 5:

As known, directed Hamiltonian Path problem is in the class NP-complete. There does not exist a deterministic polynomial time algorithm to solve it. However, there are non-deterministic polynomial time algorithm to solve it. The following is one such

Step 1. Generate random paths in G.

Step 2. Keep only those paths that begin in V_{in} and end in V_{out}

Step 3. Keep only the paths that enter exactly in n vertices (where n is the number of vertices).

Step 4. Keep only the paths that enter each vertex at least once.

Step 5. Detect if any path survived the selection. If so the answer is "yes", otherwise it is "no".

The above algorithm is the one with which that Adleman realized the directed hamiltonian path problem in a molecular biology lab. The trick lies in the way he suggests to encode the instance of the problem. For every vertex i, in the graph G, a random strand of DNA of length 20 is synthesized and denoted with O_i. Then for every edge $(i \rightarrow j)$ in G, a strand is synthesized. The $3'$ half of O_i and the $5'$ half of O_j are concatenated as shown in the figure. In this way, when we put in a test-tube for example, copies of the nucleotides $O_{i\rightarrow j}, O_{j\rightarrow k}$ and $\overline{O_j}$ ($\overline{O_j}$ is the Watson-Crick complement of O_j, the strands will anneal as shown.

O_2	TATCGGATCG	GTATATCCGA
O_3	GCTATTCGAG	CTTAAAGCTA
O_4	GGCTAGGTAC	CAGCATGCTT
$O_{2\rightarrow 3}$	GTATATCCGA	GCTATTCGAG
$O_{3\rightarrow 4}$	CTTAAAGCTA	GGCTAGGTAC
$\overline{O}_3$	CGATAAGCTC	GAATTTCGAT

We discuss the biological realization of the non-deterministic algorithm given earlier.

Step 1. For each vertex $i \in V$ and for each edge $(i \to j) \in E$, as many copies of $\overline{O}_i$ and $O_{i\to j}$ are mixed together with ligase enzyme. So, the ligation reaction results in the formation of DNA strands that encode for the random paths through the graph.

Step 2. The molecules created in the previous step are amplified by PCR, using primers O_0 and $\overline{O}_6$. In this way, DNA molecules that encode for paths that begins at V_{in} and ends at V_{out} are selected.

Step 3. With electrophoresis, the molecules that encode paths that enter exactly n vertices are selected. In Adleman's instance it means to extract the molecules which are 140 base pairs long.

Step 4. For each vertex i of the graph, only paths that enter V_i at least once are kept. The double stranded DNAs got from step 3 is melted creating single strands and, for each i, copies of $\overline{O}_i$ are synthesized and mixed with, keeping only the molecules that anneal (they are those that contain the substrand O_i, that is, those strands encode paths that enter V_i).

The number of operations (bio operations) grows linearly with n (the number of vertices of the graph). The number of different molecules required for the computation is linear in the number m of the edges of G. Step 1 requires the synthesis of $n + m$ different Oligos (a linear quantity). Step 4 requires to repeat a procedure n times and the other three steps take constant time to be executed.

Following Adleman's experiment, in [12], a DNA experiment was described for finding a solution to another NP-complete problem, the satisfiability problem (SAT). The satisfiability problem consists of a boolean expression, the question being whether or not there is an assignment of the values, true or false, to its variables, that makes the value of the whole expression true. In [2] and [4] molecular programs were given for breaking the U.S. Government's Data Encryption Standard (DES). DES encrypts 64 bit messages and uses a 56 bit key. Breaking a DES means that given one (plaintext, ciphertext) pair, we can find a key which maps the plaintext to the ciphertext. A conventional attack on DES would need to perform an exhaustive search through all the 2^{56} DES keys, which at the rate of 100,000 operations per second, would take 10,000 years. In contrast, it was estimated that DES could be broken by using molecular computation in about 4 months of laboratory work. Later on, DNA algorithm were proposed for graph connectivity problem and knapsack problem [3], road colouring problem [10], matrix multiplication [13], addition [8].

The problems mentioned above, show that molecular computation has the potential to outperform the existing computer. One of the reasons is that, the operations molecular biology currently provides, can be used to organize massive parallel searches. It is estimated that DNA computing could yield tremendous advantages from the point of view of speed, energy efficiency and economic information storing. For example, in Adleman's experiment the number of operations per second could be approximately 1.2×10^{18}. This is approximately 1,200,000 times faster than the fastest super computer.

4 About the possibility of DNA computers

DNA solutions for NP-complete problems demonstrate that molecules hide interesting computational properties. The potential advantages of DNA computing versus electronic computing are clear in the case of problems like HPP, SAT and DES. But, we should not forget that these are only particular instances of a problem solved by means of molecular biology, such results would not be so interesting if a DNA computer did not exist. This give rise to two questions. Firstly, is the DNA model of computation computationally complete in the sense that the action of any computable function (or equivalently, the computation of any Turing machine) can be carried out by DNA manipulation? Secondly, does there exist a universal DNA system i.e., a system that given the encoding of a computable function as an input, can simulate the action of that function for any argument?

Since 1936, the standard accepted model of universal computation has been the Turing machine (TM) introduced in [16]. The Church-Turing thesis, the prevailing paradigm in computer science, states that the power of Turing machines is the highest level of computability. One of the main reasons that Church-Turing thesis is widely accepted is that very diverse alternate formalizations of the class of effective procedures have all turned out to be equivalent to Turing machine formalization.

A computer is a programmable machine, able to execute any program it receives. From a theoretical point of view, this corresponds to the notion of a universal Turing machine. Consider the Turing machine M, which starts working with a string w written on its tape and reaches a final state or not (and then halts), depending on whether or not $w \in L(M)$. A Turing machine can be encoded as a string of symbols over a suitable alphabet. Let the code be denoted by code (M). Imagine a Turing machine M_u which starts working from a string which contains both $w \in T^*$ and code (M) for a given Turing machine M and stops in a final state if and only if $w \in L(M)$. Such a machine is called universal Turing machine, with a code of a particular TM is written on the tape of the universal TM, together with a string to be dealt with by the particular machine. This idea was mooted by Turing himself in [16].

Analogous to the computers (as of today), the code of a Turing machine is its program, the strings to be recognized are the input data and the universal Turing machine is the computer itself.

At this stage, it is better to know an important distinction between the terms: computational completeness and universality. Given a class C of computability models, we say that C is computationally complete if the devices in C can characterize the power of Turing machine (or any other type of equivalent devices). That is, by computationally completeness, we mean that given a Turing machine M, we can find an element C_1 in C such that C_1 is equivalent with M. Universality is an internal property of C and it means the existence of a fixed element of C which is able to simulate any given element of C. Thus, important for any theory which attempts to provide general models of computability are the completeness and universality.

So far quite a few and diverse theoretical models have been proposed for DNA-based computing [14]. While models have been based on different ideas and principles, Watson-Crick complementarity is somehow present in a computation or derivation step in a great majority of the models. This is natural, in view of the central role of complementarity in DNA operations. A typical model of DNA computing consists of augmenting a computational aspect of complementarity with some input-output format.

There are pro's and con's for each model. A model called Splicing Systems introduced by Tom Head in [9] arises to be 'one-pot' computers with all the operations carried out in principle by enzymes.

Let us first understand the cut and paste activity carried out *in vitro* on double stranded DNAs with restriction enzymes, ligases, resembling the recombination of DNA as it takes place *in vivo*. Consider the following DNAs

$$\begin{array}{ccc} CCCCCTCGACCCCC & \text{and} & AAAAAGCGCAAAAA \\ GGGGGAGCTGGGGG & & TTTTTCGCGTTTTT \end{array}$$

and two restriction enzymes with the recognition sites

$$\text{Taq 1}: \quad \begin{array}{l} 5'\ \mathrm{T\ C\ G\ A}\ 3' \\ 3'\ \mathrm{A\ G\ C\ T}\ 5' \end{array} \qquad \text{Sci NI}: \quad \begin{array}{l} 5'\ \mathrm{G\ C\ G\ C}\ 3' \\ 3'\ \mathrm{C\ G\ C\ G}\ 5' \end{array}$$

The effect of the enzymes on the two given DNA strands is cutting the strands containing its restriction site as a subsequence. The manner in which it cuts the strands is also specific for each enzyme. Because of cutting, the following four fragments are produced

$$\begin{array}{ll} 5'CCCCCT & \quad CGACCCCC3' \\ 3'GGGGGAGC & \quad\quad TGGGGG5' \end{array}$$

$$\begin{array}{ll} 5'AAAAAG & \quad CGCAAAAA3' \\ 3'TTTTTCGC & \quad\quad GTTTTT5' \end{array}$$

We observe that sticky end of the first fragment is complementary to the sticky end of the last one, while the same thing happens with the second and third fragments. The DNA strands with compatible ends can now be ligated by DNA ligase (we have to add the ligase), then the four fragments bind together, and we get

$$\begin{array}{ccc} CCCCCTCGCAAAAA & \text{and} & AAAAAGCGACCCCC \\ GGGGGAGCGTTTTT & & TTTTTCGCTGGGGG \end{array}$$

Thus the operation, we have to model consists of two phases:

1. Cut the sequences at well-specified sites
2. Paste the fragments with matching ends.

Now, we mathematically model the above operations as it takes place *in vivo*. The DNA strands can be thought of as the string of symbols. Given an alphabet Σ, two strings x and y over Σ and a splicing rule r then splicing x and y according to the splicing rule r consists of two steps

1. Cut x and y at certain positions determined by the splicing rule r
2. paste the resulting prefix of x with the suffix of y and the prefix of y with the suffix of x.

Using the formalism in modeling the splicing operations by Păun in [15], a splicing rule r over Σ is a word of the form $\alpha_1\#\beta_1\$\alpha_2\#\beta_2$ where $\alpha_1, \beta_1, \alpha_2, \beta_2$ are all strings over Σ and # and $ are markers not belonging to Σ.

Formally, we say that new strings z and w are obtained by splicing x and y according to the splicing rule $r = \alpha_1\#\beta_1\$\alpha_2\#\beta_2$ and we write

$$(x, y) \longrightarrow_r (z, w)$$

if and only if

$$\begin{array}{rclcrcl} x & = & x_1\alpha_1\beta_1 x'_1 & and & z & = & x_1\alpha_1\beta_2 y'_2 \\ y & = & y_2\alpha_2\beta_2 y'_2 & & w & = & y_2\alpha_2\beta_1 x'_1 \end{array}$$

for some $x_1, x'_1, y_2, y'_2 \in \Sigma^*$. The words $\alpha_1\beta_1$ and $\alpha_2\beta_2$ are called the sites of the splicing, while x and y are called the terms of the splicing. The splicing rule r determines both the sites and the position of the cutting: between α_1 and β_1 for the first term and between α_2 and β_2 for the second term. Note that the site $\alpha_1\beta_1$ can occur more than once in x while the site $\alpha_2\beta_2$ can occur more than once in y. Whenever this happens, the sites are chosen non-deterministically.

We can have a good idea of the way splicing works by applying it to simulate the addition of two positive numbers, n and m. Let $\Sigma = \{a, b, c\}$ and the splicing rule $r = a\#b\$c\#d$. When we splice $x = a^n b$ and $y = ca^m$ using the rule r, we get a^{n+m} and cb. Here a^{n+m} represents the sum of the integers n and m in the binary form.

This splicing operation is used as a basic tool for building a generative mechanism called Splicing System. Given a set of strings and a set of splicing rules, the generated language will consists of the strings obtained as follows: Starting from the set of axioms, we iteratively use the splicing rules to splice strings from the set of axioms or strings obtained in the preceding splicing steps.

This volume, includes a survey article "On the computational aspects of splicing" which discusses the different models of splicing systems along with a discussion on the existence of programmable DNA computers based on splicing.

4.1 Implementation problem

At the same time, the transition from the ideal world of mathematics to the empirical world of molecular biology laboratory is full of challenges. A rapid progress in the areas of biotechnology may surely advance the advent of DNA computers. The slow progress in biotechnology is just one of the many reasons. The main obstacles to creating a practical DNA computers still remain to be overcome. These obstacles are of two types.

1. Practical problems arising due to difficulties in dealing with large scale systems and in coping with errors. To name one, undesired annealings.

2. Theoretical difficulties concerning the versatility of DNA computers and their capacity to efficiently accommodate a wide variety of computational problems.

To have a good information about the "practical obstacles", one can refer to [11].

The remedy of much of these obstacles lies with bio-technology. Besides the effectiveness of bio-operations, one disadvantage in the implementation of DNA computation is the fact that size of the problem influences the concentration of the reactants. An exponential loss of concentration occurs even on sparse digraphs and this loss of concentration can become significant for problems not much larger than those solved by Adleman. Though these facts depend much on the bio-technology, a complementary approach is required in improving the accuracy of DNA computing with thinking on the lines of novel programming techniques: use of probabilistic algorithms and other modifications of the classical mathematical problem-solving tools. At the same, one has to bear in mind that most bio-operations rely on mainly manual handling of tubes which may prevent the large scale automatization of the process. i.e., a new thinking is required in designing DNA computers based entirely on operations that can be carried out by enzymes.

Another question, nice for discussion is, whether a DNA computer will out perform and finally replace the electronic computer in all practical applications. One of the most interesting properties of DNA molecules-from a computational point of view-is that storing data in DNA molecules allows for a very high information density, 1 bit per cubic nanometers [2]. This property gives molecular computation a great potential. Because of its potential parallelism resulting from the possibility of having upto 10^{18} bytes represented in a single test tube of DNA solution, with storage density of 10^{23} bytes per kilograms. These features make it tempting to conclude or conjecture that a DNA computer will be faster and more efficient than an electronic one, no matter what the application. Before concluding anything, we draw the attention of the readers, a nice comparison between humans and the computers by Lila Kari in [11].

In many realms of life, where the processing of large amounts of data quickly, efficiently and in parallel is involved, humans outshine the fanciest computers. Examples of such situations are our daily social encountering of persons, when the computers inside our skulls evaluate, in a split-second data felt by our sense organs about their shape, size, colour, sound, smell, posture, movement and expression. The information is then processed at lightning speed and out come the answer: friend or stranger, to smile or not to smile, to touch or not to touch. The most modern robots pale in comparison, having difficulties in processing the parallel sensory input, let alone in socializing. On the other hand, when faced with the task of adding up 1,000,000 fifty-digit numbers, an electronic computer will outperform the human brain anytime. The above comparison suggests that DNA computers may not totally replace the electronic computers but they may have a domain of their own. For example, the problems whose algorithms may be highly parallelized may find a place in the domain of the DNA-computers. The search of the "killer applications", an application for which DNA computers have specific advantages over the electronic one, is one good area to investigate.

5 Conclusion

To conclude, the status of DNA computation is in analogy with the results and the work of finding models of computation carried out in the 30's, which laid a strong foundation to the design of electronic computer. Another analogy is the state of VLSI in 1970's. In 1970's, the design and fabrication of a VLSI chip was perhaps less an engineering discipline than an art. All the problems of 70's were alleviated, in fact, those lead to a mature understanding of parallel algorithmic design principles and high performance architectures for VLSI. DNA computation suffers from difficulties similar to those suffered by VLSI in the 1970's.

In both the analogies, all the problems were settled nicely. In an optimistic way, one can surely and convincingly say that the results and models of DNA computation will lay a strong foundation leading to the design of DNA computers which is sure to strike the space in the near future.

References

[1] L. Adleman: Molecular Computation of Solutions to Combinatorial Problems. *Science* **266** (November 1994), 1021–1024.

[2] L. Adleman, P. Rothmund, S. Rowers, E. Winfree: On applying molecular computation to DES. II DIMACS workshop on DNA based computers, Princeton, 1996, 28–48.

[3] E. Baum, D. Boneh: Running dynamic programming algorithms on a DNA computer. II DIMACS workshop on DNA based computers, Princeton, 1996, 141–147.

[4] D. Bonech, R. Lipton, C. Dunworth: Breaking DES using a molecular computer. *Proc of a DIMACS workshop* Princeton, 1995, 37–66.

[5] K.L. Deninnghoff, R.W. Gatterdam: On the undecidability of splicing systems. *IJCM* **27** 1989, 133–145.

[6] K. Drlica: Understanding DNA and Gene cloning A Guide for the curious. John Wiley and Sons, New York, 1992.

[7] R. Freund, L. Kari, Gh. Paun: DNA computing based on splicing: The existence of universal computers. *Theory comput. systems* **32** (1999), 69–112.

[8] F. Guarnieri, C. Bancroft: Use of a horizontal chain reaction for DNA-based addition. II DIMACS workshop on DNA based computers, Princeton, 1996, 249–259.

[9] T. Head: Formal language theory and DNA : An analysis of the generative capacity of specific recombinant behaviors. *Bulletin of Mathematical Biology* **49** (1987), 737–759.

[10] N. Jonoska, S. Karl: Molecular computation of the road colouring problem. II DIMACS workshop on DNA based computers, Princeton, 1996, 148–158.

[11] Lila Kari: DNA Computing: arrival of biological mathematics. *The Mathematics Intelligence* **19 (2)**, 1997, 9–22.

[12] R. Lipton: DNA solution of hard computational problems. *Science* **268** 1995, 542–545.

[13] J. Oliver: Computation with DNA; matrix multiplication. II DIMACS workshop on DNA based computers, Princeton, 1996, 236–248.

[14] G. Paun, G. Rozenberg and A. Salomaa: DNA Computing New computing Paradigms. Springer, 1998.

[15] Gh. Paun: On the splicing operations. *Discrete Applied Mathematics* **70** 1996, 57–79.

[16] A.M. Turing: On computable numbers, with an application to the Entscheidungs problem. *Proceedings of London Mathematical Society* **42 (2)** 1936, 230–265.

R. Rama
Department of Mathematics
Indian Institute of Technology, Madras,
Chennai-36, India.
email: `ramar@iitm.ac.in`

Formal Language Aspects of Natural Computing– 2006, pp. 31–48

On the Computational Aspects of Splicing Operation

L. Jeganathan

1 Introduction

DNA computing is a fast growing research area concerned with the use of DNA molecules for the implementation of computational process. Nature possess marvelous computational capabilities. The researchers have found an analogy between the two process: "complex structure of a living being as the result of applying bio-molecular operations" and the computation of $f(w)$ from w using mathematical operations.

A Deoxyribose Nucleic Acid (DNA) is just a sequence of four different bases, arranged in a certain order. This arrangement depends on the organism to which the DNA belong. The sequence of DNA's determine the genetic information that dictates the type of organism and its development etc. The above bases, which are, Adenine, Thymine, Guanine, Cytosine can also be chemically synthesized and combined in strands. While the arrangements of bases *in vivo* determines the genetic information of the organism, researchers have found that, the arrangement of synthetic bases *in vitro* can be thought of as an encodement of the inputs of the mathematical problem in consideration.

As nature performs some bio-operations on the DNA's *in vivo*, bio-operations are performed on the synthetic DNA's (which is the encodement of a problem) based on the algorithm for the problem, we arrive at a DNA-strands which codes the solution of the problem. The solution of the mathematical problem can be got by decoding the DNA's got at the end of the process. Thus it has been shown that tools and techniques of bio-molecular laboratory can be effectively used to simulate and perform computational processes. Based on the above idea, Adleman thrilled the scientific community by solving Hamiltonian Path Problem using the DNA strands in his lab which were published in the seminal paper [1]. In fact, Adleman's work was the first one of its kind, through which it was demonstrated that a mathematical computation can be successfully carried out in a "wet lab". This marked the advent of the DNA computing. Though construction of computers from molecules was already suggested by R. Feynman, computations using DNA's saw the light only with Adleman's work.

Since then, 'DNA computing' has evolved into an exciting multidisciplinary research area. In fact, DNA computing can be considered as a unit of bio-molecular computation where in DNA's are the molecules used for the purpose.

Before the Adleman's paper, Tom Head in [7] had formulated a mathematical model of processing DNA's with restriction enzymes and ligases. The splicing systems formulated by Head, when no body talked about DNA computing, gave a boost to both the DNA computing and formal language theory. Though splicing system, was formulated in 1987, the area got developed more rapidly only after the "proof-of-principle" experiment for DNA computing carried out by Adleman.

This article does not claim for a latest survey article on DNA splicing, but makes a humble attempt to bring the flavour of splicing of strings and its computational power to the readers, with a short tour starting from DNA recombination to DNA computers.

The first section gives a detailed account of DNA recombination and section 2 discusses the modeling of the DNA recombination into a splicing operation. Section 3, 4 and 5 deals with the computational powers of splicing systems, iterated as well as non iterated. Section 6 discusses the extended *H* system. Section 7 discusses the various control mechanisms to enhance the power of extended *H* system to that of Turing Machine and the last section talks about the possibility of programmable DNA computers based on splicing.

1.1 *DNA* Recombination

Single stranded *DNA* molecules are linear polymers consisting of four bases, denoted by *A* (Adenine), *C* (Cytosine), *G* (Guanine) and *T* (Thymine). Every base is joined to the next by a phosphate group that is bonded through the 3′ carbon atom of the sugar portion of one base with the 5′ carbon atom of the sugar portion of the next base. This makes the 5′-phosphate group of one monomer and 3′ *OH* (hydroxyl) group of the other nucleotide available for bonding. This gives the orientation (5′ − 3′) for the *DNA*'s strands. There are two directions for the *DNA*'s, 5′ − 3′ direction and 3′ − 5′ direction. A *DNA* strand in 5′ − 3′ direction tells us that the molecule has 5′-phosphate at the left which is available for bonding and 3′-*OH* at the right which is available for bonding.

The base of one monomer interacts with the base of the other to form a hydrogen bond, which is subject to the restriction on the base pairing. *A* and *T* can pair together & *C* and *G* can pair together. No other pairings are possible. This pairing principle is called Watson - Crick Complementarity.

This hydrogen bond is much weaker than the phosphate bond. Using Watson-Crick complementarity, we can form a double-stranded *DNA* molecule from a single stranded *DNA* molecules of the opposite direction. Enzymes are protein that catalyze chemical reactions taking place in living cells. They are very specific and unique in their action. Without enzymes chemical reactions going on in living cells would be much slow. Since enzymes are so crucial, nature has created a multitude of enzymes that are very useful in processing them. There are restriction enzymes which recognize a specific sequence of *DNA* known as restriction site, cut the strands at those site in a way unique to the restricted enzymes, and produce two *DNA* fragments with either "blunt" or "sticky" ends. The behavior of the restricted enzymes (the pattern they recognize and the way they cut) is well known. There are catalogs with such information.

For instance, the restricted enzymes named TaqI, SciNI, HhaI, DpnI, BalI are characterized by the recognition sequences.

TaqI	T|C G A A G C|T	;	SciNI	G|C G C C G C|G
HhaI	G|C G C C G C|G	;	DpnI	G A|T C C T|A G
BalI	T G G|C C A A C C|G G T			

The way the enzymes cut is indicated above.

Consider the three molecules

5′	CCCCC	TCGA	CCCCC	3′
3′	GGGGG	AGCT	GGGGG	5′
5′	AAAAA	GCGC	AAAAA	
	TTTTT	CGCG	TTTTT	
5′	TTTTT	GCGC	TTTTT	
	AAAAA	CGCG	AAAAA	

When acting on the three molecules mentioned above, the three enzymes TaqI, SciNI, and HhaI will cut these molecules at the unique site occurring in them and the following six fragments are produced.
Situation I

(A)	5′	CCCCCT	(B)	CG ACCCCC	3′
	3′	GGGGGA GC		TGGGGG	5′
(C)	5′	AAAAAG	(D)	CG CAAAAA	3′
	3′	TTTTTC GC		GTTTTT	5′
(E)	5′	TTTTTG CG	(F)	CTTTTT	3′
	3′	AAAAAC		GC GAAAAA	5′

One can observe that, in all the fragments the overhangs are identical, *CG*, when reading in the 5′ to 3′ direction. But there is a crucial distinction. The free tips of all the overhangs created by TaqI and SciNI are at 5′-ends but the free tips of overhangs produced by HhaI are at the 3′ end. Various overhangs are provided by various restriction enzymes. For more information on restricted enzymes, one can consult [19].

Consider the nucleotides

AAAAAAA	GATC	AAAAAAA
TTTTTTT	CTAG	TTTTTTT
CCCCCC	TCGCCA	CCCCC
GGGGGG	AGCGGT	GGGGG

The action of DpnI and BalI on the above *DNA* strands produces the following fragments (with blunt ends)
Situation II

(α)	AAAAAAAGA	(β)	TC AAAAAAA
	TTTTTTTCT		AG TTTTTTT
(γ)	CCCCCCTCG	(δ)	CCACCCCC
	GGGGGGGC		GGTGGGGG

If a ligase is added in Situation I then the four fragments A, B, C and D can be bound together, either restoring the initial molecule or producing new molecules by recombination. A and D recombine and form a new molecule.

5′ CCCCC TCGC AAAAA
3′ GGGGG AGCG TTTTTT

The formation of these recombinant molecule is possible because of overhangs match. Like that C and B combine together with the help of a ligase and form a new molecule. The effect of a ligase enzyme, when introduced with a set of *DNA* fragments that have been produced by restriction enzymes is to bind compatible parts of fragments back into double stranded molecules by re-establishing the phosphokinase bonds.

If a ligase is added in Situation II, α and δ may be recombined and from a new molecule

AAAAAAAGACCACCCCC
TTTTTTTCTGGTGGGGG

This process of cutting the *DNA*'s using restriction enzymes and pasting the *DNA* fragments using ligase is called *DNA*-recombination.

The abstract model of this *DNA*-recombination phenomenon is the splicing system which was introduced in [7]. Thus a splicing system is a device for representing the fascinating string restructuring that takes place during the interactions of linear biopolymers in the presence of precisely specified enzymatic activities.

The abstraction of the *DNA* recombination is modeled as a splicing operation with the following reasonings

- Due to the Watson Crick complementarity, each strand of a ds*DNA* (double stranded DNA) uniquely determines the other strands. So, we can consider a ss*DNA* (single stranded DNA) (instead of ds*DNA*'s). In turn, ss*DNA* can be thought as strings with four alphabets. In a general set up, one can consider an arbitrary alphabets rather than the specific four letter alphabets of nucleotide.

- Given a string, the site of a restriction enzyme can be thought of as a triplet of strings. (u, x, v) where u is the "left context", v is the right context and the crossing x is the sticky end.

- The "way of cutting" of the enzymes is more important in the sense of determining the type of sticky ends (whether 5′ or 3′ end) which is crucial for the recombination. These two types of cutting can be viewed as two types of patterns (i.e. triples) say class I and class II. While recombining, we allow only the concatenation of fragments produced according to triples of the same class.

- As mentioned earlier, because of *DNA* recombination, there is a possibility of two new recombinant molecules as follows

$$w_1 = w_1' u_1 \mid x v_1 w_1''$$
$$w_2 = w_2' u_2 \mid x v_2 w_2''$$

  and the two triples (u_1, x, v_1) and (u_2, x, v_2). (We assume that these to triples belong to the same class). The fragments are

$w_1' u_1$	and	$x v_1 w_1''$	$\rightarrow$	cutting w_1 by (u_1, x, v_1)
$w_2' u_2$	and	$x v_2 w_2''$	$\rightarrow$	cutting w_2 by (u_2, x, v_2)

By the recombination we get strings

$$w_1' u_1 x v_2 w_2 \quad \text{and} \quad w_2' u_2 x v_1 w_1''$$

It is enough if we consider any one of the recombinant molecules, since the other new strip is obtained by a 'mirror' operation (i.e. cutting w_2 first and then w_1, instead of w_1 first).

- The length of the sites and the number of enzymes working together is not restricted. Since we are in the process of abstracting the recombination behavior, our assumption of no restriction on the site and length of enzymes stands justified as our model need not necessarily reflect the present status of Biotechnology. Thus, we are ready to formally define the splicing operation as introduced by Tom Head.

2 Splicing Operation

Let V be an alphabet. Let B and C be the finite set of tuples (or patterns) of the form (u, x, v) with $u, x, v \in V^*$. The string x of the pattern is called the crossing of the triple. Given two words $u\alpha\mu\beta v$, $p\alpha'\mu\beta' q \in V^*$ and two patterns (α, μ, β) and (α', μ, β') that have same crossing and are both in B or C, the splicing operation produces $u\alpha\mu\beta' q$ and $p\alpha'\mu\beta v$. The splicing operation was further modeled with the following reasoning.

- Instead of considering the pattern of few different classes, the pairs and triplets which belong to the same class were considered. A pair $((u_1, x, v_1), (u_2, x, v_2))$ means (u_1, x, v_1) and (u_2, x, v_2) are triples in the same classes and they are capable of producing recombinant strings.

- Instead of pairs of triples $((u_1, x, v_1), (u_2, x, v_2))$ one can consider pairs of pairs. The splicing of the strings $w_1' u_1 x v_1 w_1''$ and $w_2' u_2 x w_2''$ with respect to (u_1, x, v_1) and (u_2, x, v_2) is equivalent with the passing from $w_1 = w_1' u_1' v_1 w_1''$, $w_2 = w_2' u_2 v_2 w_2''$ with respect to (u_1, v_1), (u_2, v_2) where $u_1' = u_1 x$ and $u_2' = u_2 x$.

In this way, we arrive at a general and elegant formal splicing operation given by Păun is [11] as follows,

Let V be an alphabet. A quadruple $(u_1, u_2; u_3, u_4)$ of strings over V^*, is called a *Splicing rule*. For $x, y, z \in V^*$, we write

$$(x, y) \vdash_r z \text{ iff } \quad \begin{aligned} x &= x_1 u_1 u_2 x_2; \\ y &= y_1 u_3 u_4 y_2; \\ z &= x_1 u_1 u_4 y_2; \end{aligned}$$

for some $x_1, x_2, y_1, y_2 \in V^*$

We say that we splice x, y at the sites $u_1 u_2$, $u_3 u_4$, respectively and the result is z. Here the splicing operation is defined over the strings.

Thus in the Head's model of splicing, the patterns are pairs of triples, with a relation of a pair being reflexive and symmetric. Where as in Păun's definition of splicing operation, the patterns are pairs. But the relation among them could be, generally, any binary relation.

The Păun's model effectively linked in a very strong way the two steps of cutting and pasting. The relation between the two patterns means that when they cut, the generated strings are immediately pasted together, with no interference from strings cut from some other pattern. We just relate each pattern with only one pattern. This approach is powerful and it has often been chosen in the literature of splicing.

One more different definition in splicing was given by Pixton[18], where apart from the cutting and pasting steps, we substitute a part of the pattern before pasting the strings originated by the cut. The rule r is of the form $(\alpha, \alpha'; \beta)$. Such a rule r may be applied to a pair of strings that contain the sites α and α' as substrings as follows. The first string is cut before α, the second is cut before α', and the left segment of the first is attached to the right segment of the second with an intervening copy of β. If $\xi\alpha\eta$ and $\xi'\alpha'\eta'$ are the two strings, the effect of applying the rule r to the strings is $\xi\beta\eta'$.

We return back to the splicing operation. Having such an operation, we can design a language generating mechanism in the usual way; take an alphabet, take a set of words over these alphabet (axioms) and take a set of splicing rules. Starting from the axioms, we get a set of words, which forms the language of splicing. We take the splicing operation as given by Păun for our discussion; i.e. we extend the splicing operation to languages. Before that, we formally record the splicing operation.

Definition 1. Consider an alphabet V and two special symbols, # and \$. A splicing rule (over V) is a string of the form $v = u_1\#u_2\$u_3\#u_4$ where u_1, u_2, u_3, $u_4 \in V^*$. Given such a rule r, and x, y, $z \in V^*$, we write $(x, y) \vdash_r z$ iff $x = x_1u_1u_2x_2$, $y = y_1y_3y_4y_2$, $z = x_1u_1u_4y_2$ for some $x_1, x_2, y_1, y_2 \in V^*$.

3 *H*-Scheme / Splicing Scheme (Non-iterated Splicing)

A pair of $\sigma = (V, R)$ where V is an alphabet and $R \subseteq V^*\#V^*\$V^*\#V^*$ is a set of splicing rules, is called an H-scheme. Note that, R can be infinite and moreover it is itself a set of strings, hence a language. Then we can consider the complexity of of R, i.e. its place in the Chomsky hierarchy. If $R \in FL$, for a given family FL of languages, then we say that H-scheme σ is of the FL type, i.e. if R as a language is regular, Then the H-scheme is called a regular splicing type.

For a given H-scheme $\sigma = (V, R)$ and a language $L \subseteq V^*$, we define

$$\sigma(L) = \{z \in V^* | (x, y) \vdash_r z, \text{ for some } x, y \in L, \ r \in R\}$$

$\sigma(L)$ is the language obtained by a one step splicing of any pair of strings from L with respect to any rule from R.

Given two families FL_1, FL_2 we denote

$$S_1(FL_1, FL_2) = \{\sigma(L) | L \in FL_1, \sigma = (V, R), R \in FL_2\}$$

Here the suffix '1' of 'S' indicates that we are considering only one string of the two possible strings obtained by recombination (1-splicing). If we consider both the strings (2-splicing), then 'S' shall be suffixed with '2'. In this way, we interpret the suffixes '1' and '2' of the symbols H, EH and σ, wherever they occur. If no suffix is present, the suffix is taken to be 1. The symbol $\vdash$ refers 1-splicing and $\models$ refers to 2-splicing.

The following table synthesizes the known results about families $S(F_1 F_2)$ with F_1, F_2 in the Chomsky hierarchy. At the intersection of the row ranked with F_1 with the column ranked with F_2, there appear either the family $S(F_1, F_2)$ (when it is precisely identified) or two families F_3, F_4 indicating the relation $F_3 \subset S(F_1, F_2) \subset F_4$

Table 1: Power of $S(FL_1, FL_2)$

	FIN	REG	LIN	CF	CS	RE
FIN	FIN	FIN	FIN	FIN	FIN	FIN
REG	REG	REG	REG, LIN	REG, CF	REG, RE	REG, RE
LIN	LIN, CF	LIN, CF	RE	RE	RE	RE
CF	CF	CF	RE	RE	RE	RE
CS	RE	RE	RE	RE	RE	RE
RE	RE	RE	RE	RE	RE	RE

In fact, all families $S(FL_1, FL_2)$ characterise families in the Chomsky hierarchy with the exceptions $S(REG, FL_2)$ with $FL_2 \in \{LIN, CF, CS, RE\}$ and $S(LIN, FL_2)$ with $FL_2 \in \{FIN, REG\}$ which are strictly intermediate between the families in the Chomsky hierarchy. The non-iterated splicing discussed can be of mathematical interest. But for purpose of computability as well as, in view of the real phenomena, we consider the iterated splicing.

4 *H*-Scheme (Iterated Splicing)

When some restriction enzymes and a ligase are present in a test tube, they do not stop acting after one cut and paste operation, but they act iteratively.

For a H-scheme $\sigma = (V, R)$, and a language $L \subseteq V^*$, we define

$$\sigma^0(L) = L$$
$$\sigma^{i+1}(L) = \sigma^i(L) \cup \sigma(\sigma^i(L)),\ i \geq 0$$
$$\sigma^*(L) = \bigcup_{i \geq 0} \sigma^i(L)$$

Thus $\sigma^*(L)$ is the closure of L under this splicing (using the rules) in σ, i.e. the smallest language which contain L and is closed under the usual rules from σ.

Given two families FL_1, FL_2 we define

$$H_1(FL_1, FL_2) = \{\sigma_1^*(L) | L \in FL_1 \text{ and } \sigma = (V, R) \text{ with } R \in FL_2\}$$

$H_1(FL_1, FL_2)$ is for iterated splicing where as $S_1(FL_1, FL_2)$ is for non-iterated splicing.

Example 1. *For $V = \{a, b\}$, $L = \{abba\}$, let R be the set of the two splicing rules*

$$\{r_1 : a\#b\$b\#ba$$
$$r_2 : b\#a\$b\#ba\}$$

By using the first rule, we get

$$(a \mid bba,\ ab \mid ba) \vdash_{r_1} aba$$

By splicing the strings aba, abba using the rule r_1, we get the same string aba. By iteratively using the second rule, we get strings of an increasing length

$$(ab \mid a,\ ab \mid ba) \vdash_{r_2} abba$$
$$(abb \mid a,\ ab \mid ba) \vdash_{r_2} abbba$$

In general;

$$(ab^n \mid a,\ ab \mid ba) \vdash_{r_2} ab^{n+1}a,\ n \geq 1$$

Thus for the splicing scheme

$$\sigma = (V, R),\ \text{and } L = \{abba\}$$

$$\sigma(L) = \{aba, abbba\}$$
$$\sigma^0(L) = \{abba\}$$
$$\sigma^1(L) = \{aba, abba, abbba\}$$
$$\sigma^i(L) = \{ab^n a | 1 \leq n \leq i+2\},\ i \geq 1$$
$$\sigma^*(L) = \{ab^n a | n \geq 1\}$$

Table 2: Power of the $H(FL_1, FL_2)$

	FIN	REG	LIN	CF	CS	RE
FIN	FIN, REG	FIN, RE	FIN, RE	FIN, RE	FIN, RE	FIN, RE
REG	REG	REG, RE	REG, RE	REG, RE	REG, RE	REG, RE
LIN	LIN, CF	LIN, RE	LIN, RE	LIN, RE	LIN, RE	LIN, RE
CF	CF	CF, RE	CF, RE	CF, RE	CF, RE	CF, RE
CS	CS, RE	CS, RE	CS, RE	CS, RE	CS, RE	CS, RE
RE	RE	RE	RE	RE	RE	RE

Two basic results in this area are of crucial importance for *DNA* computing based on splicing.

- Regularity preserving lemma: $H(REG, FIN) \subseteq REG$.
- Basic universality lemma: Every language $L \in RE$, $L \subseteq T^*$, can be written in the form $L = L' \cap T^*$ for some $L' \in H(FIN, REG)$.

Regularity preserving lemma implies that these systems generate a restricted subclass of the class regular languages [2, 3, 6, 18]. The iterated splicing with respect to regular set of rules already lead from the finite language to non context sensitive language. In fact for all FL_2 containing the regular languages, the intersection of languages in $H(FL_1, FL_2)$ with regular languages of the form V^* characterize RE. All the families $H(FL_1, FL_2)$, $FL_1 \neq RE$ have a surprising limitation. When FL_1' is the smallest family among those considered here which strictly includes FL_1, there are languages in FL_1' which are not in $H(FL_1, FL_2)$ for all FL_2, including $FL_2 = RE$.

5 *H*-Systems (Splicing System)

H-system is a triple $\nu = (V, L, R)$, where V is the alphabet, $L \subseteq V^*$. In fact, the underlying splicing scheme is $\sigma = (V, R)$, $L(\nu) = \sigma_1^*(L)$. Note that, we allow the components L and R to be infinite, which contrasts with the usual custom of defining grammar: a finite mechanism generating possibly infinite language.

In fact, the splicing systems are renamed as *H*-systems (*H* refers Head, who introduced this). This renaming, in one way, helps to prevent the confusion of the splicing operations, as understood here, with that of *RNA* splicing.

6 Extended *H*-System

Following the model of Chomsky grammars, we can also consider auxiliary symbols, which can be used during the generation process, but are not allowed to appear in the words generated. In this way we obtain the notion of an extended *H*-system (Extended splicing system). This extended *H*-system were introduced by Păun [11].

An extended H system is a quadruple $\nu = (V, T, A, R)$ where V is an alphabet, $T \subseteq V$, $A \subseteq V^*$ and $R \subseteq V^*\#V^*\$V^*\#V^*$, and \$, # are special symbols not in V. We call T as the terminal alphabet. A is the set of axioms, R is the set of splicing rules. The underlying H scheme is $\sigma = (V, R)$. The language generated by ν is defined by

$$L(\nu) = \sigma^*(A) \cap T^*.$$

For two family of languages, FL_1, FL_2, we denote by $EH(FL_1, FL_2)$, the family of languages $L(\nu)$ generated by extended H systems $\nu = (V, T, A, R)$ with $A \in FL_1$, $R \in FL_2$.

Example 2. *Consider the extended H system*

$$\nu = (\{a, b, c\}, \{a\}, \{a^n b, ca^m\}, \{a\#b\$c\#a\})$$

n and m are two given positive integers.

We apply the splicing rule a#b\$c#a to the axioms $a^n b$, ca^m,

$$(a^n \mid b,\ c \mid a^m) \vdash (a^{n+m}, cb)$$

none of the words obtained can be spliced.

$$\sigma^0(A) = \{a^n b, ca^m\}$$
$$\sigma^1(A) = \sigma^0(A) \cup \{a^{n+m}, cb\}$$
$$\sigma^i(A) = \sigma^1(A),\ i \geq 2$$

$$\therefore \sigma^*(A) = \sigma^0(A) \cup \sigma^1(A) = \{a^n b, ca^m, a^{n+m}, cb\}$$
$$L(\nu) = \{a^{n+m}\}$$

The above extended H system ν 'computes' the addition of the integers n and m.

We can have a more simple splicing rule for the above language. That is,

$$\nu' = (\{a, b, c\}, \{a\}, \{a^n b, ca^n\}, \{\#b\$c\#\} \text{ and } L(\nu) = L(\nu').$$

In general, H-systems, extended or not, turned out to be a very powerful generative mechanism.

Table 3: Power of $EH(FL_1, FL_2)$

	FIN	REG	LIN	CF	CS	RE
FIN	REG	RE	RE	RE	RE	RE
REG	REG	RE	RE	RE	RE	RE
LIN	LIN, CF	RE	RE	RE	RE	RE
CF	CF	RE	RE	RE	RE	RE
CS	RE	RE	RE	RE	RE	RE
RE	RE	RE	RE	RE	RE	RE

Two fundamental results on EH systems are Regularity Lemma:

$$EH(FIN, FIN) = REG$$
$$EH(REG, FIN) = REG$$

Basic Universality Lemma: $EH(FIN, REG) = RE$

The regularity lemma was first proved in [3]. A direct finite automata proof is given in [11]. A result much more general than the regularity lemma is given in [18]. The proof of the basic universality lemma can be had in [16].

We observe that using a finite extended H-system, we can obtain a characterization of regular languages i.e., the power of such devices stops at the level of finite automata. From the table, it is clear that we need a context-sensitive set of axioms in order to characterize RE using a finite set of splicing rules. Instead, if we take regular set of splicing rules, we could easily get the full power of Turing machine(type-0 grammars in Chomsky hierarchy). But then, there might arise a situation wherein we have to deal with an "infinite splicing rules which might lead to infinite computers". Thus, we have two choices to investigate, either we deal with "infinite computers" or be satisfied with the computation to the level of finite automata. Of course, it might not be realistic to deal with "infinite computers". In case, we go for a finite splicing system, we cannot dream of universal DNA computers, since there is no finite automaton which is universal for the class of finite automata. The reasons are

1. We cannot encode the way of using the finite automaton in terms of a finite automaton (for this, we have to remember the symbols in the input string without marking and this cannot be done with a finite set of states).

2. We cannot codify the states of any given FA in such a way that finite set of states of the Universal automaton can handle them. (This is because, an arbitrarily large set of states will lead to arbitrarily long codes which cannot be handled by the finite set of states).

Thus, we are left with the only choice of finite splicing systems. Then arises the question: Can we at least regulate the work of these systems by adding a supplementary control on the splicing operations or on any other components of the system?.

In the next section, we consider some extended systems with the work controlled by certain mechanisms. Such controlled EH Systems can be clustered into three main categories based on the controlling mechanisms. They are

Regulated rewriting-like control

- extended H systems with permitting contexts
- extended H systems with forbidding contexts
- extended H systems with local targets
- extended H systems with a global target
- programmed extended H systems
- extended H systems with double splicing

Other control mechanisms

- extended H systems using multisets
- evolving extended H systems
- extended H systems with a fitness mapping

Distributed architectures

- communicating distributed H systems
- two-level distributed H systems
- time-varying distributed H systems.

This list is not exhaustive. Here, we fully discuss the controlled EH systems with regulated rewriting-like control and one system each for the other two categories.

7 Controlled Extended H-System

7.1 Extended H systems with permitting context

An extended H system with permitted contexts is a quadruple $v = (V, T, A, R)$ where V, T, A are as usual, R is a set of triples (rules with permitting contexts)of the form $p = (r; C_1, C_2)$ where $r = u_1\#u_2\$u_3\#u_4$ is a splicing rule over V and C_1 and C_2 are finite subsets of V^+. To such a triple p, we associate the word

$$\rho(p) = u_1\#u_2\$u_3\#u_4@w_1 \text{ and } \ldots \text{ and } w_k@v_1 \text{ and} \ldots \text{and } v_m$$

where

$$C_1 = \{w_1, \ldots, w_k\}$$
$$C_2 = \{v_1, \ldots, v_m\}, k, m \geq 0.$$

where @, \$ are new symbols.

$$\rho(R) = \{\rho(p) | p \in R\}$$

For $x, y, z, w \in v^*$ and $p \in R$, we define $(x, y) \vdash_p (z, w)$ iff $(x, y) \vdash_r (z, w)$, every word contained in C_1 appears as a subword in x, and every word contained in C_2 appears as a subword y. the language generated by v is defined as $L(v) = \sigma^*(A) \cap T^*$. The family of languages $L(v)$, for $v = (V, T, A, R)$, $A \in F_1$ and $\rho(R) \in F_2$ is denoted by $EH[F_1, pF_2]$. Instead of F_1, we put $[k]$ when systems with at most k axioms are used. Here the words over V in the sets C_1, C_2 control the use of splicing rules.

From a biochemical point of view, the permitting contexts can be interpreted as a catalysts or promoters, favouring the splicing by the associated splicing rule.

Since the controlled EH systems are intended to be theoretical models of DNA computers based on splicing, a more realistic 2-splicing operation (where in we take into account both the possible strings obtained by recombination) is considered. A more refined version of EH systems with permitting contexts can be seen in [12]. EH systems with permitting contexts was introduced in [5].

The result which shows the computational completeness of EH with permitting contexts is

Theorem 1. *$EH([1], pFIN) = EH(F_1, pF_2) = RE$, for all families F_1, F_2 such that $FIN \subseteq F_1$, $F_2 \subseteq RE$.*

The above result is proved in [5].

7.2 Extended *H*-systems with Forbidding contexts

Instead of controlling the applicability of a splicing rule by using permitting contexts, i.e., by checking the occurrence of specific symbols in the underlying words, we can also control the applicability of a splicing rule by using forbidding contexts i.e., by forbidding the occurrence of specific symbols in the underlying word. In the biochemical point of view, forbidding contexts can be interpreted as inhibitors suppressing the associated rule.

The definition of EH systems with forbidding contexts goes similar to EH-systems with permitting context but for the permitting control. The set of triples is of the form $p = (r; D_1, D_2)$, $r = u_1\#u_2\$u_3\#u_4$ is a splicing rule over V, D_1, D_2 are finite subsets of V^+.

For $x \in V^+$ and $p \in R$ as above, we define $(x, y) \vdash_p (z, w)$ iff $(x, y) \vdash_r (z, w)$, no word contained in D_1 appears as a subword in x and no word contained in D_2 as a subword in y. Language generated by v is defined as usual, the family of languages L(v), for $v = (V, T, A, R)$ with $A \in F_1$ and $\rho(R) \in F_2$is denoted by $EH(F_1, fF_2).EH([k], fF_2)$ carries the meaning as in that of permitting context. The result justifying the computational completeness with forbidding context:

Theorem 2. *$EH([1], fFIN) = EH(F_1, fF_2) = RE$, for all families F_1, F_2 such that $FIN \subseteq F_1 \subseteq RE$, $FIN \subseteq F_2 \subseteq RE$.*

The details can be had in [5].

7.3 *EH* systems with target languages

One bio-chemical aspect, encountered *in vivo* is that: nature selects the off springs of the evolutionary process by not allowing the perpetuation of 'unsuitable' forms of life. As a part of evolutionary process, mutations and recombinations are made to improve the off-spring's. Such speculations are modeled in the splicing frame-work by considering a sort of hypothesis language. This is somewhat analogous to grammar systems with hypothesis languages in grammar systems theory.

7.4 *EH* systems with local targets and global target

An *EH* system with local target is a construct $\nu = (V, T, A, R)$, where V is an alphabet; $T \subseteq V$, A is a finite language over V. R is a finite set of pairs $p = (r, Q_p)$, where $r = u_1\#u_2\$u_3\#u_4$ is a splicing rule over V and Q_p is a regular language over V.

For $x, y, z, w \in V^*$ and $p = (r, Q_p)$ in R, we write

$$(x, y) \vdash_p (z, w) \text{ iff } (x, y) \vdash_r (z, w), z, w \in Q_p.$$

i.e., the results of the splicing with respect to r belong to Q_p. For every p, we have a target Q_p. The target languages are considered in [10]. As usual, we denote the family of languages of the extended H systems with local targets $EH(FIN, ltFIN)$. $EH([n], lt[m])$, n, $m \geq 1$, denotes the family of languages with at most n axioms and splicing rules of radius at most m. (radius of a splicing rule = Max.$\{|x| \mid x = u_i, 1 \leq i \leq 4$, for some rule $u_1\#u_2\$u_3\#u_4\}$).

Instead of a target (a local one) for each p, we can have a common target for all the p's. As a result, we have an extended H system with a global target. Then the system is $\nu = (V, T, A, R', Q)$ with R' consisting of the splicing rules in R, wherein, all the other components of the system remain the same but for the replacement of lt by gt.

Theorem 3. $RE = EH([1], gt[2]) = EH([1], lt[2])$

The above result shows that we can characterize the family of RE by *EH* systems with local as well as global targets.

7.5 Programmed *EH* systems

So far, we were controlling splicing operations based on the contextual conditions. Any rule can be used at any time, but only for splicing strings fulfilling certain conditions. However, the splicing rules themselves can change from one step to the next one, influenced by the very strings to which they are applied (or by the currently available set of strings). i.e., we speak of the mechanism on the application of the rules. In such a systems, not only the splicing rules act on the strings, but strings also influence the splicing rules. This was first studied in [13].

A programmed extended H systems is a construct $\nu = (V, T, A, R, next)$ where V is an alphabet, $T \subseteq V$, A is a finite language over V (axioms), R is a finite set of splicing rules over V, and $next : R \to P(R)$, the power set of R. The language generated by ν is defined by

$$L(\nu) = (A \cup \sigma(A) \cup \rho(A)) \cap T^*$$

where $\sigma = (V, R)$ is the underlying H-scheme of ν and

$$\begin{aligned} \rho(A) = \{w \in V^* | \quad & \text{there is a sequence of splicings of the form} \\ & (x_1, y_1) \models_{r_1} (x_2, y'_2), \\ & (x_2, y_2) \models_{r_2} (x_3, y'_3), \ldots \\ & (x_k, y_k) \models_{r_k} (x_{k+1}, y_{k+1} \ldots, k \geq 2, \\ & r_i \in next(r_{i-1}), 2 \leq i \leq k+1, x_1, y_1 \in A, 2 \leq i \leq k, w = x_{k+1}\} \end{aligned}$$

$L(v)$ contains all terminal strings in A, those strings which can be obtained by one splicing starting from strings in A or by several splicings with the following properties: one starts from two axioms; at each subsequent step, one splices the first of the two strings produced at the previous step with one axiom; the rules used at consecutive steps are related by the *next* mapping.

As usual we denote $EH([n], pr[m]), n, m \geq 1$ or $EH(FIN, FIN)$, the family of languages generated by the programmed EH systems.

Theorem 4. $RE = EH([1], pr[2])$

The computational completeness of the programmed EH system is proved in [13].

7.6 *EH*-systems based on double splicing

In this variant of EH-systems, the work of an H-system proceeds in a couple of steps; the two strings obtained after a splicing immediately enters a second splicing. The rules in the two steps are not prescribed or dependent in any way to each other.

The EH systems based on double splicing is a construct $v = (V, T, A, R)$ with A and R as finite sets. For x, y, w, $z \in V^*$ and r_1, $r_2 \in R$, we write $(x, y) \models_{r_1, r_2} (w, z)$ iff $(x, y) \models_{r_1} (u, v)$ and $(u, v) \models_{r_2} (w, Z)$, u, $v \in V^*$. $\sigma_d(L)$ is the splicing language got by double splicing the strings of L using R. For $L \subseteq V^*$, we define $\sigma_d(L) = \{w \mid (x, y) \models_{r_1, r_2} (w, z)$ or $(x, y) \models_{r_1, r_2} (z, w)$ for $x, y \in L$, r_1, $r_2 \in R\}$. $\sigma_d{}^*(L) = \cup_{i \geq 0} \sigma_d{}^i(L)$ and $L_d(v) = \sigma_d{}^*(A) \cap T^*$. As usual, $EH(FIN, dFIN)$ denotes the families of languages $L_d(v)$ generated by the EH systems based on double splicing. When there is a bound on the splicing rules, we write $EH(FIN, d[k])$, where the splicing rules in R are of radius at most k. As expected, the extended H-systems using the double splicing operation are equal in power to type-0 grammar.

Theorem 5. $RE = EH_2(FIN, d[2])$.

This system was introduced in [15].

7.7 Extended *H*-systems with multisets

So far, in our discussions on splicing, after splicing two strings x, y and obtaining the two possible new strings z, w, we may use again x or y as a term of splicing. these strings are not consumed by the splicing. Further, we may splice x or y with z or w. Thus, we may splice strings from one 'generation' with strings from another 'generation'. Also these new strings z and w are to appear in an arbitrary number of copies of each. We follow an assumption that if a string is available then arbitrarily many copies of it are available. Usually, a large number of copies of each string is used whenever a string is used.

How ever the existence of several copies of each string raises the difficult problem of controlling the splicing so as to prevent "wrong" operations. A possibility to avoid this difficulty is to use at least some of the strings in a specified number of copies, and to keep track of these numbers during the work of these systems. This leads to multisets (sets with multiplicities associated with their elements). This multiset concept was introduced in [4] and later improved in [5].

A multiset over a set X is a function $M :\rightarrow N \cup \{\infty\}$; $M(x)$ is the number of copies of $x \in X$ in the multiset M. The set $\{w \in X \mid M(w) > 0\}$is called the support of M and is denoted by Supp(M).

An extended H system with multisets (or an extended mH system) is a construct $\nu = (V, T, A, R)$, where V, T, R are as in extended H system and A is a multiset over V^+. For such mH systems, for two multisets M_1, M_2 over V^*, we define, $M_1 \Longrightarrow_r M_2$ iff there are $x, y, z, w \in V^+$ such that

1. $M_1(x) \geq 1, (M_1 - \{(x, 1)\}(y) \geq 1$
2. $x = x_1u_1u_2x_2, y = y_1u_3u_4y_2$
   $z = x_1u_1u_4y_2, w = y_1u_3u_2x_2$
   $for\ x_1, x_2, y_1, y_2 \in V^*$, $u_1\#u_2\$u_3\#u_4 \in R$
3. $M_2 = (((M_1 - \{(x, 1)\}) - \{(y, 1)\}) \cup \{(z, 1)\}) \cup \{(w, 1)\}$.

In other way, when passing from a multiset M_1 to a multiset M_2, according to ν, the multiplicities of two elements of M_1 namely x and y is diminished by one the multiplicity of the resulting words, z and w, is augmented by one. The multiplicity of all other elements in Supp(M_1) is not changed. The obtained multiset is M_2.

The language generated by an extended mH system ν consists of all words containing only terminal symbols and whose multiplicity is at least once greater or equal to one during the work of ν. Formally,

$$L(\nu) = \{w \in T^* \mid w \in supp(M) \text{ for some } M \text{ such that } A \Rightarrow_r^* M\}.$$

The family of languages generated by the extended mH systems are denoted by $EH(mFIN, FIN)$.$EH(m[n], [m])$ denotes the family of languages generated by the extended mH systems with cardinality (supp(A))$\leq$ n and Radius(R)$\leq$ m. The computational completeness of mH systems are given by the result

Theorem 6. *$REG = EH(m[1], FIN) \subseteq EH(m[2], FIN) = EH(mF_1, F_2) = RE$ for all families F_1, F_2 such that $FIN \subseteq F_1 \subseteq RE$, $FIN \subseteq F_2 \subseteq RE$*

It is proved in [5].

7.8 Time-Varying distributed *H*-Systems

One of the main drawbacks of the controlled *H*-system is that we need several splicing rules. Each rules correspond to two variations. Each enzyme needs specific temperature, acidity, salinity and other reaction conditions. For a nice implementation of the system in the lab, we have to change all the reaction conditions. Of course, this decreases the efficiency of the computation. One way to increase the power of *H* systems with finite sets of axioms is to organise the computation in a distributed way. We separate the parts of the model which are able to work independently, preferably in parallel, arranging a way of computation between these parts and synthesizing the result of the computation from the partial result produced by the mentioned parts.

One well investigated class is that of time varying *H* system. The passing from a set of rules to another one is now specified in a cycle.

In the sense of Bio-Chemistry, these models start from the assumption that the splicing rules are based on the enzymes which are essentially based on environmental conditions. At any moment, only a subset of the set of all available rules are active. If the environment changes periodically, then the active enzyme class changes periodically.

A time varying distributed H system of degree $n \geq 1$ is a construct

$$\Gamma = (V, T, A, R_1, R_2, \ldots, R_n),$$

where V, T, A are defined as usual. R_i are the finite set of splicing rules over V, $1 \leq i \leq n$. The sets R_i are called the components of the system.

At each moment $k = nj + i$ for $j \geq 0$, $1 \leq i \leq n$, the component R_i is used for splicing the currently available strings.

We define

$$L_1 = A$$
$$L_{k+1} = \sigma_2^{(i)}(L_k) \text{ for } i \equiv k \mod n,\ k \geq 1$$

where $\sigma(i) = (V, R_i)$, $1 \leq i \leq n$.

Language generated by Γ is defined by

$$L(\Gamma) = \left(\bigcup_{k\geq 1} L_k\right) \bigcap T^*.$$

We denote by VDH_n, $n \geq 1$, the family of languages generated by time varying distributed H-systems of degree at most n and VDH_* is used to denote the family of all languages of this type.

Time varying distributed H system was first introduced in [12]. The result $RE = VDH_7$ is proved in [15]. This result has been improved to $VDH_1 = RE$ in [9].

8 Programmable DNA computers based on splicing

In the previous sections, we discussed that the extended H-systems with finite axioms and finite splicing rules supplemented by some controls are computationally complete, in the sense of Turing machines. However, such results are not enough in order to provide programmable computability models (computers) based on the splicing operations. To this aim, universal H systems (of the considered types), should be found in the sense introduced by Turing. A Turing machine TM_u is universal if all its components are fixed and given a specific Turing machine TM_0, we can run TM_0 as a program in the following sense.

Assume that starting from word w, the machine TM_0 produces a word w, the machine TM_0 produces a word w'. There is a word w_0, the code of TM_0, such that starting from w_0w, the machine TM_u produces the word w' and this is true for all pairs (w, w') having w_0 fixed. Thus w_0 is the 'program' of the algorithm described by TM_0. TM_u receives w_0 as well as some input w and behaves exactly as TM_0 when working on the same data. The similarity with the actual computers is apparent.

Thus the universal H-system is a system with all the components but one fixed, able to behave as any given splicing system v when a code v is introduced in the set of axioms of the universal system.

Definition 2. [Universal H-systems] Given an alphabet T and two families of languages F_1, F_2, a construct $v_u = (V_u, T, A_u, R_u)$, where V_u is an alphabet, $A_u \subseteq V_u{}^*$, $A_u \in F_1$ $R_u \subseteq V_u{}^*\#V_u{}^*\$V_u{}^*\#V_u*$, $R_u \in F_2$ is said to be a universal H-system of type (F_1, F_2), if for every $v = (V, T, A, R)$ of type (F_1, F_2), there is a language A_v such that $A_u \cup A_v \in F_1$ and $L(v) = L(v'_u)$ where $v'_u = (V_u, T, A_u \cup A_v, R_u)$.

Surprisingly enough, universal H systems exist for all the types of extended H-systems so far investigated, in the sense of programmable DNA computation based on splicing. The following two results are important in the sense of programmable DNA computation based on splicing.

Theorem 7. *Given an extended H-system v (of course, with finite axioms and finite splicing rules) of any one of the types discussed(such as permitting context etc). Let V be the total alphabet and T be the terminal alphabet. We can construct an extended H-system v′ of the same type with v, with total alphabet $T \cup \{c_1, c_2\}$ and the terminal alphabet T such that $L(v) = L(v')$.*

The interpretation of the above result is that the number of auxiliary symbols used when passing from a type-0 grammar to an extended H-system of the mentioned types can be significantly decreased. Two such symbols are enough.

We mention the main result from the point of view of DNA computability. we mention the result for mH systems. But similar results hold true, *mutatis mutandis*, for all types of H systems discussed above.

Theorem 8. *For every given alphabet T, there exists an extended mH system of type $(m[2], FIN)$ which is universal for the class of extended mH system with the terminal alphabet.*

The interpretation of these results, from the point of view of DNA computing, is that, there exist programmable universal DNA computers which are based on the splicing operations. Here, the stress is much on the theoretical character of the results. i.e., it is proved that the DNA computers, programmable and universal, are mathematically possible. However, a discussion on the feasibility of such DNA computers goes beyond the realms of mathematics.

9 Conclusion

As stated earlier, though this survey article does not claim to be an exhaustive one, this article may provide a good insight and a nice motivation for the computational aspects of the splicing operation. Since the DNA molecules are present in the form of a circular sequences, the theory of splicing is extended to circular strings (A circular string is a sequence $x = a_1a_2 \ldots a_n$ for $a_i \in V$, $1 \leq i \leq n$ with the assumption that a_1 follows a_n). Splicing of circular strings was introduced in [8] and studied in [17, 18, 20, 21].

References

[1] L.M. Adleman: Molecular Computation of solution to combinatorial problem. Science, 226 (1994), 1021–1024.

[2] K. Culik II and T. Harju: The regularity of splicing systems and DNA. Lecture Notes in Computer Science 372 (1989), 63–69.

[3] K. Culik II and T. Harju: Splicing semigroups of dominoes and DNA. Discrete. Appl. Math., 31 (1991), 261–267.

[4] K.L. Denninghoff and R.W. Gatterdam: On the undecidability of splicing systems. International journal of computer mathematics, 27 (1989), 133–145.

[5] R. Freund, L. Kari, Gh.Păun: DNA Computing based on splicing. Theory of computing systems, 32, 69–112 (1999).

[6] R.W. Gatterdam: Splicing systems and regularity. Intnl. jr. Computer mathematics, 31 (1989), 63–67.

[7] T. Head: Formal Language theory and DNA: An analysis of the generative capacity of specific recombinant behaviours. Bulletin of Mathematical Biology, 49 (1987), 737–759.

[8] T. Head: Splicing schemes and DNA. in Lindenmayer Systems; Impacts on Theoretical Computer Science and Developmental Biology (G. Rozenberg, A. Salomaa, eds) Springer, Berlin (1992), 371–383.

[9] M. Margenstein and Y. Rogozhin: Time-varying distributed *H*-system of degree 2. Proc. 4th International Meeting on DNA based computers, Pheladelphia (June 1998).

[10] Gh. Păun: Splicing systems with targets are computationally complete. Inform. processing letters, 59 (1996), 129–133.

[11] Gh. Păun, G.Rozenberg and A.Salomaa: Computing by splicing. Theoretical Computer Science, 168(2)(1996), 321–336.

[12] Gh. Păn: DNA Computing: Distributed splicing systems. Structures in Logic and Computer Science. A selection of essays in honor of A Ehrenfeuchet Lecture Notes in Computer Science, 1261, Springer Berlin (1997), 351–370.

[13] Gh. Păun, G. Rozenberg and A. Salomaa: Computing by splicing, programmed and evolving systems. IEEE Inter. conf. on evol. computing, Indianapolis, (1997) 273-277.

[14] Gh. Păun, G. Rozenberg and A. Salomaa: DNA computing. Springer, 1998.

[15] Gh. Păn: DNA Computing based on splicing: Universality results. Proceedings of the second intern. colloq. universal machines and computations, Metz, 1998, 1, 67–91.

[16] Gh. Păun: On Time-varying H-systems. Bull. EATCS 67 (1999) 157–164.

[17] D. Pixton: Linear and circular splicing system. IEEE Conf. on Intelligence in neural and biological systems, Heindon - Washington (1995), 181–188.

[18] D. Pixton: Regularity of splicing languages, Disc. Applied Maths. 69 (1996), 101–124.

[19] R.J. Roberts, S.M. Lim and R.S. Loyd(eds): Nucleases, 2ed Cold Spring Harbor Laboratory Press, 1993.

[20] R. Siromoney, K.G. Subramanian and V.R. Dare: Circular DNA and splicing systems. Parallel Image Analysis, Lecture Notes in Computer Science, 654, Springer, Berlin (1992), 260–273.

[21] T. Yokomori, S. Kobayashi and C. Ferretti: On the power of circular splicing systems and DNA computability, Report CSIM-95-01, Univ. of Electro-Comm., Chofu, Tokyo, 1995 and IEEE intern. conf on Evolutionary computing, Indianapolis, 1997, 219–224.

L. Jeganathan
Department of Mathematics
Indian Institute of Technology Madras
Chennai - 600 036, India

Formal Language Aspects of Natural Computing– 2006, pp. 49–61

A Glimpse of Membrane Computing

Kamala Krithivasan

1 Introduction

'Natural Computing' is an area which is pursued with interest in recent times. 'Cellular computing' is a part of Natural Computing. One model of cellular computing is membrane computing (P Systems) [61] initiated by Gh. Paun in 1998, aiming at devising a computing model inspired by the structure and functioning of living cells. It is a parallel and distributed model of computing. In this paper we give a glimpse of this area. This paper is not attempting to survey all the results in this area which has grown enormously in the least few years.

In the basic model of P systems, one considers a *membrane structure* consisting of several cell-like membranes which are hierarchically embedded in a main membrane, called the *skin membrane*. Membranes with no other membranes embedded in them are called *elementary*. The membranes delimit *regions*, where we place *objects*. The objects evolve in nondeterministic maximally parallel manner according to given evolution rules. The objects can also be described by strings and then they are processed by string operations. Starting from an initial configuration, identified by membrane structure and objects in all regions, and using evolution rules, we get a *computation*. We consider a computation complete when it *halts*, i.e., no further rule can be applied. The result of the halting computation is defined on the basis of the objects we obtained in a specified membrane or expelled from the skin membrane.

We may look at membranes to correspond to regions in cells and objects to chemical compounds and use rules to describe the evolution of those chemical compounds.

2 Definitions

In this section we give some basic definitions.

2.1 Membrane Structure

The basic structural ingredient of P systems is a membrane structure. We formalize a membrane structure by means of well-formed parenthesized expressions, strings of correctly matching parentheses, placed in a unique pair of matching parentheses. Each pair of matching parentheses corresponds to a membrane. We now present the definition for membrane structure as given in [13].

Let us consider the language MS over the alphabet $\{[,]\}$, whose strings are recursively defined as follows:

- $[\,] \in MS$;
- if $\mu_1, \ldots, \mu_n \in MS, n \geq 1$, then $[\mu_1, \ldots, \mu_n] \in MS$;
- nothing else in MS.

For each pair $x, y \in MS$, $x \sim y$ if and only if $x = \mu_1\mu_2\mu_3\mu_4$, $y = \mu_1\mu_3\mu_2\mu_4$, where $\mu_1\mu_4 \in MS$ and $\mu_2, \mu_3 \in MS$. We denote by $\sim^*$ the reflexive and transitive closure of the relation $\sim$. Let $\overline{MS}$ denote the set of equivalence classes of MS with respect to this relation. The elements of $\overline{MS}$ are called *membrane structures*.

The number of membranes in a membrane structure μ is called the *degree* of μ and denoted by $deg(\mu)$.

The *depth* of a membrane structure μ, denoted by $dep(\mu)$, is defined recursively as follows:

- if $\mu = [\,]$, then $dep(\mu) = 1$;
- if $\mu = [\mu_1 \ldots \mu_n]$, then $dep(\mu) = \max\{dep(\mu_i) \mid 1 \leq i \leq n\} + 1$ for some $\mu_1, \ldots, \mu_n \in MS$.

Graphically a membrane structure is represented by a Venn diagram without intersection and with a unique superset as shown in Figure 1. A membrane structure corresponding to the Venn diagram is $\mu = [_1[_2]_2[_3]_3[_4[_5]_5[_6[_8]_8[_9]_9]_6[_7]_7]_4]_1$.

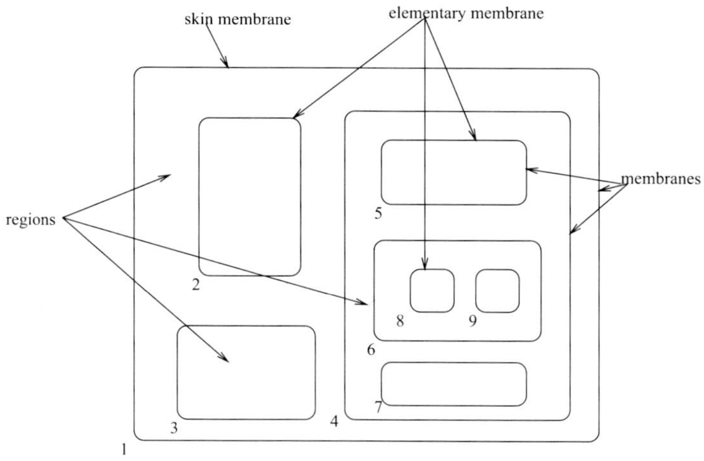

Figure 1: A membrane structure

2.2 Super-Cell

In a cell many objects can be considered as being *atomic*. So we follow this notion to define this computing model. We now present the definition of *super-cell* as given in [13].

Let V be a finite set, whose elements are called *objects*, and μ be a membrane structure of degree n, $n \geq 1$, as defined in the previous subsection. Assume that the membranes are labeled in a one-to-one manner with the numbers $\{1, \ldots, n\}$. Then we define super-cell as a construct which contains both membrane structure μ and multiset $M_i : V \rightarrow N$ associated with each region i of μ, $1 \leq i \leq n$.

Any multiset M_i mentioned above can be empty and each object can appear in several regions, in several copies in each of them. The multiset corresponding to a region of a super-cell is called the *contents* of the region. The total multiplicities of the elements in an elementary membrane m is called the *size* of membrane m and is denoted by *size*(m). We can define *degree*, *depth*, and *region* of a super-cell similar to a membrane structure.

We are now ready to formally define this computing model.

2.3 Formal Definition

A *P system* [13] is a super-cell provided with evolution rules for its objects and with a designated output region.

Definition 1 (P System). A *P system* of degree n, $n \geq 1$, is a construct

$$\Pi = (V, T, C, \mu, w_1, \ldots, w_n, (R_1, \rho_1), \ldots, (R_n, \rho_n), i_0),$$

where:

- V is the total alphabet of the system;
- $T \subseteq V$ is the terminal alphabet;
- $C \subseteq V$, $C \cap T = \emptyset$, is the set of catalysts;
- μ is a membrane structure consisting of n membranes, with the membranes and the regions labeled in a one-to-one manner with the elements from the set $\{1, \ldots, n\}$;
- w_i, $1 \leq i \leq n$, are strings representing multisets over V associated with the regions $1, \ldots, n$ of μ;
- R_i, $1 \leq i \leq n$, are finite sets of *evolution rules* over V associated with the regions $1, \ldots, n$ of μ; ρ_i is a partial order relation over R_i, specifying a *priority* relation among the rules of R_i;
- i_0 specifies the *output* membrane of Π, if it is a number between 1 and n. If it is equal to ∞, output is read in the outer region.

An evolution rule is of the form $u \to v$, with $u \in V^+$ and $v = v'$ or $v = v'\delta$, where v' is a string over $(V \times \{here, out\}) \cup (V \times \{in_j \mid 1 \leq j \leq n\})$, and δ is a special symbol not in V. The length of the string u is called the *radius* of the rule $u \to v$. When presenting the evolution rules, the indication *here* is in general omitted.

The use of priority relations can be interpreted either in a *strong* sense or in a *weak* sense. In the *strong* sense: if a rule with higher priority is used, no rule of a lower priority can be used even if the two rules do not compete for the same objects. In the *weak* sense: a rule is always used when objects exist which were not used by a rule of a higher priority.

If Π contains rules of radius greater than *one*, we say that Π is a system *with cooperation*, otherwise, it is a *non-cooperative* system. A particular class of cooperating systems is that of *catalytic* systems: the only rules of radius greater than *one* are of the form $ca \to cv$, where $c \in C$, $a \in (V - C)$, and v contains no catalyst; moreover, no other evolution rule contains catalysts (there is no rule of the form $c \to v$ or $a \to v_1cv_2$ for $c \in C$). Catalysts will not evolve but they help other objects to evolve.

The membrane structure μ and the multisets $w_1, \ldots, w_n$ in Π constitute the *initial configuration* of the system. We can pass from one configuration to another by using the evolution rules. This is done in parallel: all objects, from all membranes, which can be the subject of local evolution rules, should evolve simultaneously. A rule can be used only when there are copies of the objects whose evolution it describes and which are not *consumed* by the rules of higher priority and, moreover, there is no rule of a higher priority, irrespective of which objects it involves, which is applicable at the same step. Therefore, all objects to which a rule can be applied must be the subject of a rule application. All objects in u are consumed by using the rule $u \rightarrow v$. The result of using the rule, in region t, is determined by v. If an object appears in v in the form $(a, here)$, it will remain in the same region. If an object appears in v in the form (a, out), a will exit membrane t and become an element of the region that surrounds membrane t. In this way, it is possible that an object leaves the system: if it goes outside the skin membrane, it never comes back. If an object appears in the form (a, in_q), a will be sent to membrane q, provided membrane q is adjacent to membrane t. If (a, in_q) appears in v and membrane q is not one of the membranes delimiting *from below* the region t, the application of the rule is not allowed.

Instead of in_i, we can consider weak communication by introducing a new target notation "*in*" [64]. The meaning of (a, in) is that one occurrence of the symbol a is sent to any one of the directly inner membranes, chosen nondeterministically.

If the symbol δ appears in v, membrane t is dissolved and at the same time the set of rules R_t (and its associated priority relations) is removed. The multiset from membrane t is added to the multiset associated with the region which is directly external to membrane t.

All these operations are performed in parallel, for all possible rules $u \rightarrow v$, for all occurrences of multisets u in the regions associated with the rules, for all regions at the same time. If at the same step we have (a, in) outside a membrane t and δ inside this membrane, because of the simultaneous performance of these operations, no contradiction appears: we assume that a is introduced in membrane t at the same time when it is dissolved, thus a will remain in the region surrounding membrane t.

A sequence of transitions between configurations of a given P system Π is called a *computation* with respect to Π. A computation is *successful* if and only if it halts, i.e., there is no rule applicable to the objects present in the last configuration. The system is provided with an elementary membrane $i_0 \neq \infty$ as an output membrane in the last configuration of the computation. It is not necessary that the output membrane is an elementary one in the initial configuration. The result of a successful computation is the number of objects present in the output membrane of Π in the halting configuration. Note that the objects which are not in the output membrane are not counted. Also, it is important to note that a computation which is not halting yields no output. The set of numbers computed in this way by the system Π is denoted by $N(\Pi)$ and we say that the set $N(\Pi)$ is *generated* by Π in the *internal* mode.

A variant of P systems was considered in [63], where no output membrane is specified, i.e., $i_0 = \infty$, so that we have to observe the system from outside and collect the objects ejected from the skin membrane. The result of a halting computation can be taken in two ways. In the first case, we count the number of objects sent out of the system Π during the computation. The set of numbers computed in this way by the system Π is denoted by $N(\Pi)$ and we say that the set $N(\Pi)$ is *generated* by Π in the *external* mode. In the second case, we collect the objects sent out of the system Π in the order in which they are ejected. Using these objects we form a string. When several objects are ejected at the same time, any permutation of them is considered. In this way, a string or a set of strings is associated with each computation, i.e., a language is associated with the system Π. We denote by $L(\Pi)$ the language computed by Π in the way described above and say that it is *generated* by Π

in the *external* mode. As in the case of internal mode, here also a computation which is not halting yields no output.

We can use a P system for solving three types of tasks: as a *generative device* - start from an initial configuration and collect all sets of natural numbers describing the multiplicities of objects from an output membrane or ejected from the membrane system at the end of a successful computation; as a *computing device* - start with some input placed in an initial configuration and read the output at the end of a successful computation; and as a *decidability device* - introduce a problem in an initial configuration, in a specific encoding, and look for the answer after a specified number of steps.

3 Variants of P systems

Several variants of P systems have been proposed and shown to be computationally universal. Here we list some of them. Details can be found in [1]. So far, we consider *P systems with symbol-objects* where objects are being *atomic*, but in a cell some objects, such as e.g., DNA molecules, have a structure, which can be described by a string. This leads one to consider *P systems with string-objects* [24, 52, 49, 61]. Depending upon the way of processing string-objects, P systems are further divided into *rewriting P systems* and *splicing P systems*. Rewriting P systems, where rewriting rules are used for processing string-objects, were introduced in [61] and thoroughly studied in [10, 11, 21, 26, 37, 38, 39, 48, 52, 58, 69, 74, 75]. String-objects in a P system can also be processed by using the *splicing* operation introduced in [32] as a formal model of the DNA recombination under the influence of *restriction enzymes* and *ligases*. If we consider such splicing rules for processing string-objects, we obtain splicing P systems. Splicing P systems were introduced in [61] and investigated in [25, 29, 60, 66]. In P systems with symbol-objects, we work with multisets and the result of a computation is a set of natural numbers; in the case of P systems with string-objects, we work with sets of strings and the result of a computation is a set of strings. Combining these two ideas, a new variant called *P systems with worm-objects* was introduced in [14] and studied in [51].

Motivated from the fundamental idea of *vectors* used in *gene cloning* [12] and the fact that the *carrier proteins* assist molecules to pass through membranes [2], a variant called *P systems with carriers* was introduced independently in [41, 54] and studied in [33].

By considering the biological phenomenon of *trans-membrane transport* of couples of chemicals, a variant called *P systems with Symport/Antiport* was introduced in [59] and investigated in [27, 30, 35, 50, 28]. The term *symport* indicates the passage of two chemicals, together through a membrane, in the same direction. If two chemicals pass simultaneously through a membrane in opposite direction, it is called *antiport*.

One of the most important ways of moving chemicals through bio-membranes is based on concentration differences between adjacent regions. This idea was formalized in [22]. Another variant [65] which has a good motivation is that of systems with *energy accounting*: integer numbers are associated with rules, expressing *quanta of energy* produced or consumed when applying these rules; at each step, we can use a combination of rules only if the total energy in each region is positive; one starts with zero units of energy in the system, and the energy remaining from an application of a step is passed on to the next step; however, if the total energy within a membrane is bigger than the threshold associated (in advance) with the membrane, the membrane is dissolved (and all its energy consumed in this way).

Combining the ideas of inter-cellular communication and the way the neurons cooperate and process impulses in the complex net established by *synapses* [3], a new mathematical model called *tissue P systems* was proposed in [53]. Tissue P systems consist of several *cells*, related by protein channels. Each cell has a state from a given finite set and can process multisets of objects, represented by symbols from a given alphabet.

An important theorem from [73] says that each deterministic P system with symbol-objects can be simulated by a deterministic Turing machine with a polynomial slowdown. This means that by using such systems we cannot solve exponential problems in linear time. In order to improve the computational performance of P systems, different variants have been proposed. One such variant, motivated from the biological phenomenon of membrane division, is *P systems with active membranes* [58, 62, 73]. Since one membrane can divide into number of membranes, we get exponential space which can be useful for solving NP-complete problems. The operation of moving a membrane, together with its contents, through another membrane, corresponds to the biological operations of *endocytosis* and *fagocytosis*, was considered in [5] in the framework of P systems with active membranes, as a way to avoid the division of membranes under the influence of inner membranes. Another way for obtaining exponential space is by using the possibility of creating new membranes. Such variant was considered in [42, 68, 36].

Relationships between P systems and distributed computing were investigated in [57]. A recent approach to this question, of possible usefulness of P systems as models of distributed computing, can be found in [18]. The purpose is to prove that *P systems can become a primary model for distributed computing, particularly for message-passing algorithms*. Algorithms are presented for several communication tasks, such as skin membrane broadcast, generalized broadcast, convergecast, leader election, fault-tolerant consensus, etc. Message complexity and time complexity are evaluated for the membrane algorithms corresponding to these problems.

An important direction of research concerns the normal forms, mainly with respect to the membrane structure: it is expected that certain structures are easier to handle than others; in particular, the *lyposomes* are more frequently arranged in a *linear way* than in a *branched way*. Normal form theorems concerning the membrane structure can be found in [56, 64]. Normal forms with respect to the number and the form of rules present in each membrane, for the case of rewriting P systems, were given in [74, 75].

Several papers have considered the fundamental topic of implementing P systems on an electronic computer, either on the existing media or on a purposely designed architecture. The first type of approach appears in [4, 6, 7, 8, 19, 47, 43, 72], while the second one appears in [71].

The approach considered in the study of P systems is clearly motivated from a mathematical point of view, not only because it is natural to model the cell computational behavior, but also because the new computing model has several intrinsically interesting features, like, the use of multisets, the inherent parallelism, the possibility of devising computations which can solve NP-complete problems in polynomial time, etc.

4 Recent Trends

4.1 P Automata

P automata are defined based on the concept of considering P system as accepting devices. In P automata, each membrane has a state like Tissue P systems and computation starts at an initial state and ends in a final state. In [20] one model is considered where rules are used in a sequential manner. In [46] a variant is considered where rules can be applied in maximal mode. Computational universality is shown. Ph.D., thesis by Ludek Cienciala considers several aspects of P automata [40].

4.2 PX systems

Combining the concept of X machines [23] and P systems PX systems are defined and studied in [9].

4.3 On Hierarchies

A open problem was posed in [67]: Is there a non universal model of a membrane computing system which induces an infinite hierarchy on the number of membranes. This was answered in the affirmative by O.H. Ibarra by considering RCPS systems in [34]. Another paper about infinite hierarchy by Krishna and Rama is appearing in this volume.

4.4 Forbidding and Enforcing Conditions

Forbidding and enforcing in membrane computing with biochemical motivation are considered in [15]. Two approaches are presented there.

4.5 Array Objects

P systems with array objects, which are rectangular array of symbols are considered in [16, 17]. In [16] array rewriting rules are considered whereas in [17] contextual rules are considered.

4.6 Complexity

Deterministic and nondeterministic complexity classes of P systems are studied in [55]. Some descriptional complexity issues of rewriting P systems are studied in [44].

4.7 Conformon P systems

Conformon P systems are considered by P. Frisco et al [31]. A separate article by Frisco appears in this volume.

4.8 Computational Capability

Peter Sosik has compared the generative power of some variants of P systems with complexity classes like PSPACE-complete and also shown that some variants may have super Turing computational power [70].

5 Conclusion

Membrane Computing has emerged as a big area of research in the recent years. So far about 14 Ph. D. theses and about 700 publications have come out in this area. The entire details can be found in the webpage [1]. We have not attempted to give details of all recent results about P systems. We have mentioned in this paper only about those results which we personally found very interesting. A number of open problems remain. A list is given in the webpage. With more and more people getting into this area, the field is growing very fast. Currently there is also a lot of interest in using this for practical problems.

References

[1] `http://psystems.disco.unimib.it`.

[2] B. Alberts, D. Bray, J. Lewis, M. Raff, K. Roberts, and J.D. Watson. *Molecular Biology of the Cell.* Garland Publishing, New York, 3rd edition, 1994.

[3] M.A. Arbib. *Brains, Machines, and Mathematics.* Springer, Berlin, 2nd edition, 1987.

[4] F. Arroyo, A.V. Baranda, J. Castellanos, C. Luengo, and L.F. Mingo. Structures and bio-language to simulate transition P systems on digital computers. In C.S. Calude, Gh. Păun, G. Rozenberg, and A. Salomaa, editors, *Multiset Processing. Mathematical, Computer Science, and Molecular Computing Points of View*, volume 2235 of *Lecture Notes in Computer Science*, pages 1–16. Springer-Verlag, Berlin, 2001.

[5] A. Atanasiu and C. Martín-Vide. Recursive calculus with membranes. *Fundamenta Informaticae*, 49(1-3):45–59, 2002.

[6] A.V. Baranda, J. Castellanos, F. Arroyo, and R. Gonzalo. Towards an electronic implementation of membrane computing: A formal description of nondeterministic evolution in transition P systems. In N. Jonoska and N.C. Seeman, editors, *DNA Computing. 7th International Workshop on DNA-Based Computers, DNA7, Tampa, Florida, USA, June 10-13, 2001. Revised Papers*, volume 2340 of *Lecture Notes in Computer Science*, pages 350–359. Springer-Verlag, Berlin, 2002.

[7] A.V. Baranda, J. Castellanos, F. Arroyo, and C. Luengo. Bio-language for computing with membranes. In J. Kelemen and P. Sosik, editors, *Advances in Artificial Life. 6th European Conference, ECAL 2001, Prague, Czech Republic, September 10-14, 2001. Proceedings*, volume 2159 of *Lecture Notes in Artificial Intelligence*, pages 176–185. Springer-Verlag, Berlin, 2001.

[8] A.V. Baranda, J. Castellanos, R. Gonzalo, F. Arroyo, and L.F. Mingo. Data structures for implementing transition P systems in silico. *Romanian Journal of Information Science and Technology*, 4(1-2):21–32, 2001.

[9] F. Bernardini, M. Gheorghe and M. Holcombe. PX systems = P Systems + X machines. Natural Computing, 2, 201–213, 2003.

[10] D. Besozzi, C. Ferretti, G. Mauri, and C. Zandron. Parallel rewriting P systems with deadlock. In M. Hagiya and A. Ohuchi, editors, *DNA Computing. 8th International Workshop on DNA Based Computers, DNA8, Sapporo, Japan, June 10-13, 2002. Revised Papers*, volume 2568 of *Lecture Notes in Computer Science*, pages 302–314. Springer-Verlag, Berlin, 2003.

[11] P. Bottoni, A. Labella, C. Martín-Vide, and Gh. Păun. Rewriting P systems with conditional communication. In W. Brauer, H. Ehrig, J. Karhumäki, and A. Salomaa, editors, *Formal and Natural Computing. Essays Dedicated to Grzegorz Rozenberg*, volume 2300 of *Lecture Notes in Computer Science*, pages 325–353. Springer-Verlag, Berlin, 2002.

[12] T.A. Brown. *Gene Cloning. An Introduction*. Chapman and Hall, London, 1996.

[13] C.S. Calude and Gh. Păun. *Computing with Cells and Atoms*. Taylor and Francis, London, 2001.

[14] J. Castellanos, Gh. Păun, and A. Rodríguez-Patón. P systems with worm-objects. In *IEEE 7th International Conference on String Processing and Information Retrieval, SPIRE2000*, pages 64–74, La Coruna, Spain, 2000.

[15] M. Cavaheie and N. Jonoska. Forbidding and enforcing in membrane computing. Natural Computing, 2, 215–228.

[16] R. Ceterchi, M. Mutyam, Gh. Paun and K.G. Subramanian. Array rewriting P systems. Natural Computing, 2(3), 229–249, 2003.

[17] K.S. Dersanambika and K. Krithivasan. Contextual Array P systems. IJCM, 81, 955–969, 2004.

[18] G. Ciobanu, R. Desai, and A. Kumar. Membrane systems and distributed computing. In Gh. Păun, G. Rozenberg, A. Salomaa, and C. Zandron, editors, *Membrane Computing. International Workshop, WMC-CdeA 2002, Curtea de Argeş, Romania, August 19-23, 2002. Revised Papers*, volume 2597 of *Lecture Notes in Computer Science*, pages 187–202. Springer-Verlag, Berlin, 2003.

[19] G. Ciobanu and D. Paraschiv. P system software simulator. *Fundamenta Informaticae*, 49(1-3):61–66, 2002.

[20] E. Csuhaj-Varj*ú* and G. Vazil. P automata or purely communicating accepting P systems. In Gh. P*ă*un, G. Rozenberg, A. Salomaa, and C. Zandron, editors, *Membrane Computing. International Workshop, WMC-CdeA 2002, Curtea de Argeş, Romania, August 19-23, 2002. Revised Papers*, volume 2597 of *Lecture Notes in Computer Science*, pages 187–202. Springer-Verlag, Berlin, 2003.

[21] J. Dassow and Gh. Păun. On the power of membrane computing. *Journal of Universal Computer Science*, 5(2):33–49, 1999.

[22] J. Dassow and Gh. Păun. P systems with communication based on concentration. *Acta Cybernetica*, 15(1):9–14, 2001.

[23] S. Eilenberg. Automata, Languages and Machines. Academic Press, 1974.

[24] C. Ferretti, G. Mauri, Gh. Păun, and C. Zandron. On three variants of rewriting P systems. In C. Martín-Vide and Gh. Păun, editors, *Pre-Proceedings of Workshop on Membrane Computing, WMC-CdeA 2001*, pages 63–76, Curtea de Argeş, Romania, August 20-25, 2001. Technical Report 17/01 of Research Group on Mathematical Linguistics, Rovira i Virgili University, Tarragona, Spain, 2001, and *Theoretical Computer Science* (article in Press).

[25] R. Freund. Generalized P systems with splicing and cutting/recombination. *Grammars*, 2(3):189–199, 1999.

[26] R. Freund, C. Martín-Vide, and Gh. Păun. From regulated rewriting to computing with membranes: Collapsing hierarchies. Submitted, 2001.

[27] R. Freund and M. Oswald. A short note on analysing P systems with antiport rules. *Bulletin of the EATCS*, 78:231–236, 2002.

[28] R. Freund and A. Păun. Membrane systems with symport/antiport rules: Universality results. In Gh. Păun, G. Rozenberg, A. Salomaa, and C. Zandron, editors, *Membrane Computing. International Workshop, WMC-CdeA 2002, Curtea de Argeş, Romania, August 19-23, 2002. Revised Papers*, volume 2597 of *Lecture Notes in Computer Science*, pages 270–287. Springer-Verlag, Berlin, 2003.

[29] P. Frisco. On two variants of splicing P systems. *Romanian Journal of Information Science and Technology*, 4(1-2):89–100, 2001.

[30] P. Frisco and H.J. Hoogeboom. Simulating counter automata by P systems with symport/antiport. In Gh. Păun, G. Rozenberg, A. Salomaa, and C. Zandron, editors, *Membrane Computing. International Workshop, WMC-CdeA 2002, Curtea de Argeş, Romania, August 19-23, 2002. Revised Papers*, volume 2597 of *Lecture Notes in Computer Science*, pages 288–301. Springer-Verlag, Berlin, 2003.

[31] P. Frisco and R.T. Gibson. A simulator for conform on P systems. Proceedings of WMC 6, 355–372, 2005.

[32] T. Head. Formal language theory and DNA: An analysis of the generative capacity of specific recombinant behaviors. *Bulletin of Mathematical Biology*, 49:737–759, 1987.

[33] H.J. Hoogeboom. Carriers and counter: P systems with carriers vs. (blind) counter automata. In M. Ito and M. Toyama, editors, *Proceedings of 6th International Conference on Developments in Language Theory*, pages 255–268, Kyoto, Japan, September 18-21, 2002.

[34] O.H. Ibarra. On membrane hierarchy in P systems. Theoretical Comp. Sci. 334, 115–120, 2005.

[35] M. Ionescu, C. Martín-Vide, A. Păun, and Gh. Păun. Unexpected universality results for three classes of P systems with symport/antiport. In M. Hagiya and A. Ohuchi, editors, *DNA Computing. 8th International Workshop on DNA Based Computers, DNA8, Sapporo, Japan, June 10-13, 2002. Revised Papers*, volume 2568 of *Lecture Notes in Computer Science*, pages 281–290. Springer-Verlag, Berlin, 2002.

[36] M. Ito, C. Martín-Vide, and Gh. Păun. A characterization of Parikh sets of ET0L languages in terms of P systems. In M. Ito, Gh. Păun, and S. Yu, editors, *Words, Semigroups, and Transducers*, pages 239–254. World Scientific, Singapore, 2001.

[37] S. N. Krishna and R. Rama. On the power of P systems with sequential/parallel rewriting. *International Journal of Computer Mathematics*, 76(1-2):317–330, 2000.

[38] S. N. Krishna and R. Rama. Note on parallel rewriting in P systems. *Bulletin of the EATCS*, 73:147–151, 2001.

[39] S. N. Krishna and R. Rama. P systems with replicated rewriting. *Journal of Automata, Languages and Combinatorics*, 6(3):345–350, 2001.

[40] Ludek Cienciala. P automata. Ph.D. Thesis. University of Ostrava, Ostrava, Czech Republic, 2005.

[41] M. Madhu and K. Krithivasan. Inter-membrane communication in P systems. *Romanian Journal of Information Science and Technology*, 3(4):335–352, 2000.

[42] M. Madhu and K. Krithivasan. P systems with membrane creation: Universality and efficiency. In M. Margenstern and Y. Rogozhin, editors, *Machines, Computations, and Universality. 3rd International Conference, MCU 2001, Chisinau, Moldava, May 23-27, 2001. Proceedings*, volume 2055 of *Lecture Notes in Computer Science*, pages 276–287. Springer-Verlag, Berlin, 2001.

[43] M. Madhu, V.S. Murthy, and K. Krithivasan. Hardware realization of P systems with carriers. In M. Hagiya and A. Ohuchi, editors, *Proceedings of 8th International Meeting on DNA based Computers*, Sapporo, Japan, 2002 (Poster).

[44] M. Madhu. Complexity issues in rewriting P systems. Proceedings of fourth international workshop on descriptional complexity of Formal Systems, London, Ontario, Canada, 21–24, 2002.

[45] M. Madhu. Studies of P systems as a model of cellular computing, Ph.D. Thesis. IIT Madras, Chennai, 2003.

[46] M. Madhu and K. Krithivasan. On a class of P automata. International Journal of Computer Mathematics, 80(9), 1111-1120, 2003.

[47] M. Malita. Membrane computing in Prolog. In C.S. Calude, M.J. Dinneen, and Gh. Păun, editors, *Pre-proceedings of Workshop on Multiset Processing*, pages 159–175, Curtea de Argeş, Romania, 2000. TR140, CDMTCS, Auckland University.

[48] V. Manca, C. Martín-Vide, and Gh. Păun. On the power of P systems with replicated rewriting. *Journal of Automata, Languages and Combinatorics*, 6(3):359–374, 2001.

[49] M. Margenstern, C. Martín-Vide, and Gh. Păun. Computing with membrane: Variants with an enhanced membrane handling. In N. Jonoska and N.C. Seeman, editors, *DNA Computing. 7th International Workshop on DNA-Based Computers, DNA7, Tampa, Florida, USA, June 10-13, 2001. Revised Papers*, volume 2340 of *Lecture Notes in Computer Science*, pages 340–349. Springer-Verlag, Berlin, 2002.

[50] C. Martín-Vide, A. Păun, and Gh. Păun. On the power of P systems with symport rules. *Journal of Universal Computer Science*, 8(2):317–331, 2002.

[51] C. Martín-Vide and Gh. Păun. Computing with membranes. One more collapsing hierarchy. *Bulletin of the EATCS*, 72:183–187, 2000.

[52] C. Martín-Vide and Gh. Păun. String objects in P systems. In *Proceedings of Workshop on Algebraic Systems, Formal Languages and Computations*, pages 161–169, RIMS Kokyuroku, Kyoto University, Kyoto, 2000.

[53] C. Martín-Vide, Gh. Păun, J. Pazos, and A. Rodríguez-Patón. Tissue P systems. *Theoretical Computer Science*, 296(2):295–326, 2003.

[54] C. Martín-Vide, Gh. Păun, and G. Rozenberg. Membrane systems with carriers. *Theoretical Computer Science*, 270(1-2):779–796, 2002.

[55] M.J. Perez Jimenez, A.R. Jimenez and F.S. Caparrini. Complexity classes in models of cellular computing with membranes. Natural Computing, 2(3), 265–285, 2003.

[56] I. Petre. A normal form for P systems. *Bulletin of the EATCS*, 67:165–172, 1999.

[57] I. Petre and L. Petre. Mobile ambients and P systems. *Journal of Universal Computer Science*, 5(9):588–598, 1999.

[58] A. Păun. On P systems with membrane division. In I. Antoniou, C.S. Calude, and M.J. Dinneen, editors, *Unconventional Models of Computation*, pages 187–201. Springer, London, 2000.

[59] A. Păun and Gh. Păun. The power of communication: P systems with symport/antiport. *New Generation Computing*, 20(3):295–306, 2002.

[60] A. Păun and M. Păun. On membrane computing based on splicing. In C. Martín-Vide and V. Mitrana, editors, *Where Mathematics, Computer Science, Linguists, and Biology Meet*, pages 409–422. Kluwer, Dordrecht, 2001.

[61] Gh. Păun. Computing with membranes. *Journal of Computer and System Sciences*, 61(1):108–143, 2000.

[62] Gh. Păun. P systems with active membranes: Attacking NP-complete problems. *Journal of Automata, Languages and Combinatorics*, 6(1):75–90, 2001.

[63] Gh. Păun, G. Rozenberg, and A. Salomaa. Membrane computing with external output. *Fundamenta Informaticae*, 41(3):259–266, 2000.

[64] Gh. Păun, Y. Sakakibara, and T. Yokomori. P systems on graphs of restricted forms. Submitted.

[65] Gh. Păun, Y. Suzuki, and H. Tanaka. P systems with energy accounting. *International Journal of Computer Mathematics*, 78(3):343–364, 2001.

[66] Gh. Păun and T. Yokomori. Membrane computing based on splicing. In E. Winfree and D. Gifford, editors, *Preliminary Proceedings of 5th International Meeting on DNA Based Computers*, pages 213–227, MIT, USA, 1999.

[67] Gh. Păun. Membrane Computing - An Introduction. Springer-Verlag. Berlin. 2002.

[68] A. Rodríguez-Patón. On the universality of P systems with membrane creation. *Bulletin of the EATCS*, 74:229–233, 2001.

[69] P. Sosik and R. Freund. String rewriting sequential P systems and regulated rewriting. In W. Kuich, G. Rozenberg, and A. Salomaa, editors, *Developments in Language Theory. 5th International Conference, DLT 2001, Vienna, Austria, July 16-21, 2001. Revised Papers*, volume 2295 of *Lecture Notes in Computer Science*, pages 379–388. Springer-Verlag, Berlin, 2002.

[70] P. Sosik and O. Valik. On evolutionary ligases of membrane systems. Proceedings of WMC 6, 79–93, 2005.

[71] Gh. Stefan. Membrane computation with cellular automata on a dedicated hardware. In C. Calude, M.J. Dinneen, and Gh. Păun, editors, *Proceedings of Worskhop on Multiset Processing*, Curtea de Argeş, Romania, August 2000.

[72] Y. Suzuki and H. Tanaka. On a LISP implementation of a class of P systems. *Romanian Journal of Information Science and Technology*, 3(2):173–186, 2000.

[73] C. Zandron, C. Ferretti, and G. Mauri. Solving NP-complete problems using P systems with active membranes. In I. Antoniou, C.S. Calude, and M.J. Dinneen, editors, *Unconventional Models of Computation*, pages 289–301. Springer, London, 2000.

[74] C. Zandron, C. Ferretti, and G. Mauri. Universality and normal forms on membrane systems. In R. Freund and A. Kelemenova, editors, *Proceedings of Workshop on Grammar Systems*, pages 61–74, Bad Ischl, Austria, July 2000.

[75] C. Zandron, C. Ferretti, and G. Mauri. Two normal forms for rewriting P systems. In M. Margenstern and Y. Rogozhin, editors, *Machines, Computations, and Universality. 3rd International Conference, MCU 2001, Chisinau, Moldava, May 23-27, 2001. Proceedings*, volume 2055 of *Lecture Notes in Computer Science*, pages 153–164. Springer-Verlag, Berlin, 2001.

Kamala Krithivasan
Department of Computer Science and Engineering
Indian Institute of Technology Madras,
Chennai - 600 036, India
`email: kamala@cs.iitm.ernet.in`

Formal Language Aspects of Natural Computing– 2006, pp. 63–76

Peptide Computing: A Survey

M. Sakthi Balan, Helmut Jürgensen and Kamala Krithivasan

Abstract

As a brief survey on peptide computing, we present some of the results from the first two related papers [12, 27] and discuss various issues concerning these results. The future direction of work in this area is also addressed.

1 Introduction

Looking for computational models beyond the silicon model is a fascinating area of research. The main motivation for looking at computational models different from the Turing model arises from the existence of problems, termed intractable, which require a large amount of run time even in supercomputers. One of the reasons for these intractable problems to take a huge amount of processing time is that there is no elegant way for attacking them. The only known method to solve these problems is through a trial-and-error method, that is, through an exhaustive search of the solution space. This exhaustive search, performed in computer, which is deterministic, requires exponential running time.

In order to perform an exhaustive search efficiently, in the exponential-size solution space, we need a model which can execute the search in parallel. That this could be achieved by computing with bio-molecules was first realized by L. Adleman [1]: he solved an instance of the Hamiltonian-path problem using only DNA strands together with some bio-operations on them. While the instance considered initially was rather small, the idea of using DNA strands has triggered a large volume of work on this area. Now DNA computing stands as an efficient model to solve at least some of the intractable problems.

The idea of using living cells and molecular complexes as potential computing components dates back to the late 1950s, when R. Feynman delivered his famous paper describing "sub-microscopic" computers [23]. Biomolecular computing is one of the computational models which uses sub-microscopic molecules – DNA and proteins.

But only after L. Adleman's pioneering work [1] in 1994 the world realized the importance of looking at bio-molecules as a computational model. DNA Computing has many advantages; it is massively parallel, computes with extremely high energy-efficiency and stores enormous quantities of information. Using massive parallelism, a DNA computer could outstrip an electronic computer on highly complex combinatorial problems – breaking encryption schemes, for example. But there is still a long way to building an efficient DNA computer. There are disadvantages too: DNA Computing is too error-prone to achieve its theoretical potential. Surveys on DNA computing can be found in [3, 21, 29, 30, 31].

Beyond computing models which exploit the biochemical properties of molecules, proposals also exist for using physical or chemical properties of molecules (see for example [2, 17, 18, 19, 20, 22]). In [4, 13, 14, 15, 16] there is a detailed description of unconventional models of computing: quantum computing, DNA computing and membrane computing.

Another computing model based on biomolecular processes, similar to but different from, DNA computing, using antibodies and their ability to recognize and bind to parts of peptide sequences is called *peptide computing*. This idea was first introduced by H. Hug and R. Schuler [27]. They provide a solution of the satisfiability problem using peptides and antibodies.

A peptide is a sequence of amino acids attached by covalent bonds called peptide bonds. A peptide consists of recognition sites, called *epitopes,* for the antibodies. A peptide can contain more than one epitope for the same or different antibodies. With each antibody, which attaches to a specific epitope, a binding power is associated, called its *affinity.* When antibodies compete for recognition sites – which may overlap in the given peptide – then the antibodies with greater affinity have higher priority.

Replacing an antibody by one with a greater affinity can be considered a computational step in a re-writing process: the antibody with smaller affinity is removed – *affinity-based removal* – and an antibody with greater affinity attaches to the epitope which became free. Antibodies can also be removed by adding excess epitopes – *epitope-based removal.* In the sequel, the term *peptide computing* refers to computational processes based on these as elementary operations.

The parallel interaction between the peptide sequences and the antibodies helps to perform exhaustive searches for combinatorially difficult problems efficiently.

There are advantages in peptide computing compared to DNA computing: at each position there can be 20 different building blocks instead of four. Moreover, while a DNA strand has only a single reverse complement of the same length, the peptide-antibody interaction provides for a far greater versatility [24, 32]:

- different antibodies can recognize different regions or structures of the same epitopes;
- the binding affinities of different antibodies recognizing the same epitope can be different.

In this paper we present a brief survey of results in peptide computing. We emphasize the issues and challenges. In the next section we focus on the two results from [27, 12]. First we outline a method for solving the satisfiability problem by peptide computing as given in [27]; then we describe how to solve the Hamiltonian-path problem as given in [12]. We also discuss some disadvantages of the approaches and propose a potential remedy. In Section 3, we describe the simulation of a Turing machine by peptide computing as given in [12]. We discuss the limitations of this simulation and outline the features of an improved simulation as proposed in [8]. The penultimate section covers some of the automaton models inspired by peptide computing and some other models for Boolean circuits and simple arithmetic operations and discuss some of the open problems in this area. In Section 5, we provide some general observations.

2 Solving NP-Complete Problems by Peptide Computing

In this section we present two results from [27] and [12]: solving the satisfiability problem and the Hamiltonian-path problem. As each of the problems is NP-complete, any problem in NP can be solved in polynomial time by peptide computing if one uses classical polynomial-time translations in pre- and post-processors. In this section, we also discuss the short-comings of the approach and potential remedies.

2.1 Satisfiability Problem

Let F be a formula in conjunctive normal form, that is F is the conjunction of finitely many clauses, C_1 to C_k, such that each clause is the disjunction of finitely many literals, that is, variables or negated variables. An assignment X attributes a truth value to every variable. A clause becomes true if it contains a variable with a truth-value of true or a negated variable with a truth-value of false. F is true under assignment X if all clauses C_1 to C_k are true under assignment X; in this case, X *satisfies* F.

A formula F is *satisfiable* if it is true under some assignment to the variables. The satisfiability problem is to decide, for a given a formula F, whether it is satisfiable and, when it is, to find an assignment which satisfies F. Let n denote the number of distinct variables used in F. Then there are $N = 2^n$ possible assignments.

Consider the following simple example. Let F consist of three clauses, C_1, C_2 and C_3, that is, $F = C_1 \wedge C_2 \wedge C_3$ where $C_1 = v_1 \vee \neg v_2$, $C_2 = \neg v_2$ and $C_3 = v_1 \vee v_2$. Every assignment gives a truth value to the variables v_1 and v_2.

The four possible assignments X_i are listed in the following table:

X_i	$X_i(v_1)$	$X_i(v_2)$
X_1	*false*	*false*
X_2	*false*	*true*
X_3	*true*	*false*
X_4	*true*	*true*

For each assignment X we prepare a peptide and antibodies binding to different overlapping epitopes as shown in Fig. 1. Each peptide, that is, each assignment X has an epitope specific to an antibody A. The epitopes for antibodies B and C consist of the same sequences in each peptide, respectively.

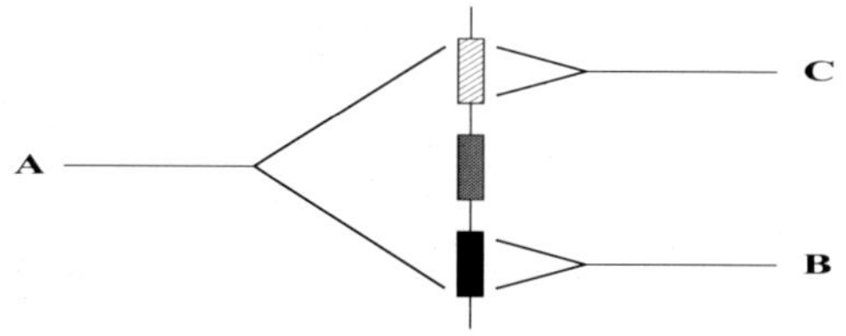

Figure 1: Binding sites of a peptide

In Fig. 1, binding sites of the antibodies A, B and C to their corresponding epitopes are shown. Antibody A recognizes all three epitopes. The antibody B and the antibody C are the same for every assignment. The antibody A is specific for each assignment. Antibody A recognizes an epitope which includes a specific part (gray) and the epitopes for B (black) and C (hatched). The binding affinity of C is greater than that of A, and the affinity of A is greater than that of B. The order of the epitopes on the peptide is arbitrary. In our example, four antibodies A_1 to A_4 are needed one for each of the assignments X_1 to X_4. Note that B and C can be present at the same time, whereas A blocks binding of B, and C blocks binding of A. To test the satisfiability of a formula $F = C_1 \wedge \cdots \wedge C_k$, we prepare sets G_1 to G_k of partial solutions as follows: G_i contains those antibodies A specific to the assignment under which C_i is true. In our example C_1 is true under assignments X_1, X_3 and X_4. C_2 is true under assignments X_1 and X_3. C_3 is true under assignments X_2, X_3 and X_4. Then we prepare $G_1 = \{A_1, A_3, A_4\}$, $G_2 = \{A_1, A_3\}$ and $G_3 = \{A_2, A_3, A_4\}$.

In the sequel, a linker is meant by a chemical which makes the binding of antibodies (which are already binding) to its epitopes a permanent one; hence those antibodies cannot be removed by adding excess epitopes nor by antibodies with high affinity.

In general, the following steps are necessary for solving the satisfiability problem:

1. Let $m = k$.
2. While $m > 0$, repeat the following steps:
   (a) The antibody set G_m is added. The antibodies A of G_m bind to their epitopes.
   (b) Antibodies B are added. Antibodies B bind to all free binding sites for B.
   (c) Antibodies C are added. Antibody C binds to all of its epitopes, since it has the highest binding affinity. Note

   that all antibodies A of set G_m are removed, whereas antibodies B remain bound to their epitopes.
   (d) Antibodies C are removed by adding epitopes for C in excess.
   (e) All remaining antibodies are attached to their epitopes by adding linker.
   (f) Let $m = m - 1$.
3. Add labeled antibody B.
4. If the presence of labeled antibodies is detected then this shows that the formula is *satisfiable*; if not, the formula is not *satisfiable*. If the formula is *satisfiable* and the peptides are addressed by some method then the exact solution can be deciphered.

In our example, when G_3 is added the antibodies A_i bind to all peptide sequences which correspond to the assignment that makes the formula *true*. Hence A_1, A_3 and A_4 bind to the peptide sequences corresponding to the assignments X_1, X_3 and X_4 respectively. When antibody B is added it binds to all peptide sequences corresponding to the assignments that makes the formula F *false*. When C is added the antibodies A_i are removed since the affinity of C is greater than that of A_i. After C is removed by adding excess epitopes for C, linker is added which attaches all remaining antibodies B to their epitopes. In this way the peptide sequences to which B is binding are made unusable for further computations as the antibody B cannot be removed from them after the linker is added. Hence, in the next iteration only the peptide sequences (i.e., assignments) for which the formula is *true*, remain active. So only the peptide sequences corresponding to X_1, X_3 and X_4 will be participating. After the last step of this iteration only the peptide sequences corresponding to X_1 and X_3 will be active as the assignment X_4 makes the clause C_2, *false*. In the last iteration G_1 is added and after the antibody C is removed only the peptide sequence corresponding to X_3 remains without any antibody binding to it. Hence when labeled antibody B is added it binds to the peptide sequence corresponding to X_3 which can then be detected. If, moreover, we can detect the exact peptide sequence to which the labeled antibody is binding then we can decipher the exact solution as X_3. Similar protocols are possible for other NP-complete search problems.

2.2 Hamiltonian-Path Problem

Let $G = (V, E)$ be a directed graph where $V = \{v_1, v_2, \ldots, v_n\}$ is the vertex set of the graph and $E = \{e_{ij} \mid v_j \text{ is adjacent to } v_i\}$ be the edge set. Without loss of generality we take v_1 as the source vertex and v_n as the end vertex. Our problem is to test whether there exists a Hamiltonian path between v_1 and v_n.

Before describing the method of solving this problem by peptide computing we define a term called *duplicated permutation* which we use in the sequel. For a finite sequence $M = m_1, m_2, \ldots, m_n$, we define the duplicated sequence of M, ΔM as

$$\Delta M = m_1, m_1, m_2, m_2, \ldots, m_n, m_n.$$

We define a duplicated permutation of a set S, $\sigma(S)$ as

$$\sigma(S) = \{\Delta m \mid m \text{ is a permutation of the set } S\}.$$

For each $v_i \in V$, choose an epitope. We denote the set of all chosen epitopes as $EP = \{ep_j \mid 1 \leq j \leq n\}$. The antibodies and peptides with epitopes from the set EP are formed as follows:

1. Each peptide has an epitope corresponding to the source vertex (ep_1) at the top and an epitope corresponding to the end vertex (ep_n) at the bottom.

2. Between these two epitopes all duplicated permutations of other epitopes corresponding to the vertices excluding the source and end vertices are formed.

3. Form antibodies A_{ij}, $1 \leq i, j \leq n$ which bind to the sites $ep_i ep_j$, provided v_j is adjacent to v_i in G.

4. Form antibodies B_{ij}, $1 \leq i, j \leq n$ which bind to the sites $ep_i ep_j$, provided v_j is not adjacent to v_i in G.

5. Form a labeled antibody C, which binds to the sites $ep_1 \sigma(\{ep_j \mid 2 \leq j \leq n-1\}) ep_n$.

6. The affinity of the antibodies B_{ij} is greater than the affinity of C.

7. The affinity of the antibody C is greater than that of A_{ij}.

The algorithm is as follows:

1. Put all the peptides formed as above into an aqueous solution.

2. Add antibodies A_{ij}, $1 \leq i, j \leq n$.

3. Add antibodies B_{ij}, $1 \leq i, j \leq n$.

4. Add labeled antibody C.

5. If the presence of labeled antibody C binding to the peptide sequences is detected then there exists a Hamiltonian path in the graph G, otherwise there exists no such path. If the peptides are addressed then we can infer the exact Hamiltonian path in the graph G.

Suppose the graph G is as in Fig. 2, then there will be 24 peptide sequences corresponding to each of the 24 possible Hamiltonian paths from the designated vertices 1 to 6. The antibody C will select the correct Hamiltonian path "132546" after the antibodies A_{ij} and B_{ij} ($1 \leq i, j \leq 6$) bind to their corresponding epitopes. For $1 \leq i, j \leq 6$, note that the antibody A_{ij} is added only if there is an edge from the vertex i to the vertex j and B_{ij} is added only if there is no edge from the vertex i to vertex j.

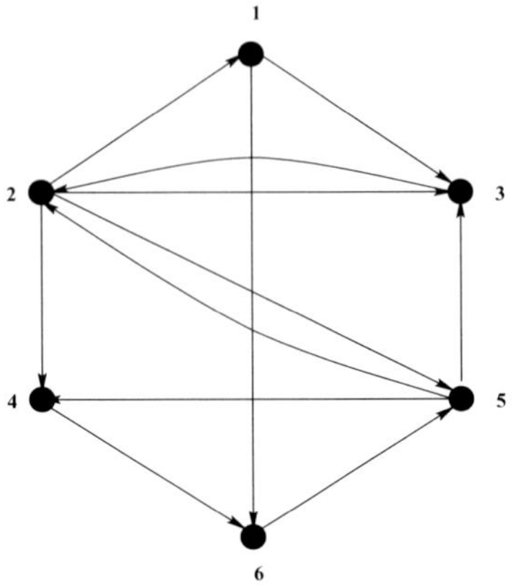

Figure 2: The graph G (from [12])

One sees that this model requires

1. $(n-2)!$ peptides;
2. peptides of length $= O(n)$;
3. $O(n^2)$ antibodies.

The number of biological steps needed to solve this problem is constant. The first step detects all possible edges in the graph in parallel. The second step detects all the invalid edges, and the third step finds the correct solution, if it exists. The fourth step is for detecting whether there exists a solution and if it exists to read the solution. For more explanations regarding this problem with an example see [12].

2.3 Discussion

The two methods for solving NP-complete problems have several drawbacks. We discuss these here.

First, both methods use an exponential number of peptide sequences: take the case of the satisfiability problem, if we have n variables, then k can be as large as 2^n. To form a clause, we could pick i variables. There are $\binom{n}{i}$ such choices. Then, for each variable, we choose whether it is negated or not. For each clause we have 2^i such choices. Then there are $2^i\binom{n}{i}$ distinct clauses involving i variables. Hence k can only be bounded (if we exclude repetitions) by

$$\sum_{i=1}^{n} 2^i \binom{n}{i}.$$

hence k can be exponential in n. This means that the preprocessing time for peptide computing is exponential in n. It will, of course be linear in k. Moreover, the satisfiability problem even needs an exponential number of antibodies. To prepare these peptide sequences and antibodies, exponential time is required. Thus the time cost is shifted from computation to preprocessing. Hence it would be important to find a method which reduces the number of peptide sequences to a feasible level.

The main reason for the exponential number of peptide sequences is that we form the peptide sequence for each potential solution of the problem in advance. One of the methods to address this issue might be an *incremental approach*. For example instead of using the method given in Section 2.2, we could build the peptide sequences incrementally; first the sequences are built with only one epitope; they are checked if they are representing valid edges in the graph by putting antibodies A and B; the invalid ones can be rejected at once and the valid ones can be used for further elongations to check the other edges in the path the sequence is representing. This method could drastically reduce the number of peptide sequences. But the number of steps which is constant for the method presented in Section 2.2 will grow linearly with respect to the number of edges. It is to be noted however that, for the method in Section 2.1, it is not obvious how this approach works or helps.

Second, the peptide sequences are problem-specific: particularly in the case of the model given in Section 2.2. Even though the peptide sequences formed for a graph having n vertices can be used for any graph with the same number of vertices, when n changes we need new peptide sequences. Hence the preprocessing time for the formation of peptide sequences is problem-dependent and a human interface is needed. Thus the method is not a general one. It would be preferable if the formation of the peptide sequences were performed as a step inside the algorithm rather than as a preprocessing step. This would allow for an automated formation of peptide sequences. But how to incorporate the formation of peptide sequences in the algorithm is an open problem. The idea will be very similar to the approach followed by Adleman [1] where DNA strands exist for each vertex and they are selected in such a way that two strands hybridize into one if they are adjacent. Hence the formation of DNA sequences representing paths in the graph is performed automatically. If the formation of peptide sequences is included in the algorithm we might be able to infer what the basic operations needed on the peptide sequences are and hence we can have a general understanding of what a peptide algorithm might look like.

3 Universality of Peptide Computing

In this section we present and discuss the simulation of a Turing machine by peptide computing as proposed in [12]. In Section 3.2 we will explain the drawbacks of this model and present some ideas as to how to overcome these problems.

3.1 Simulating a Turing Machine

In the following theorem, if M is a Turing machine then t_M and s_M denote the time and space complexity of M, respectively.

Theorem 1. [12] *Let* $M = (Q, \Sigma, \delta, q_0, F, \flat)$ *be a Turing machine. There is a simulation of* M *by peptide computing with the following properties:*

(1) There is a constant $c > 0$*, independent of* M*, such that the number of peptide-antibody*

interactions needed for the simulation of a computation step of M is no greater than c. As a consequence, the number of peptide-antibody interactions needed for the simulation of a computation of M on input $w \in \Sigma^*$ is no greater than $c \cdot t_M(w)$.

(2) *The length of the peptide sequence needed for the simulation of a computation of M on input $w \in \Sigma^*$ is in $\Theta(s_M(w))$; moreover the number of antibodies needed is in $\Theta((|Q| + |\Sigma|) \cdot s_M(w))$.*

Proof Idea [12]. Let $M = (Q, \Sigma, \delta, q_0, F, \flat)$ be a Turing machine with state set $Q = \{q_0, q_1, \ldots, q_{m-1}\}$, input alphabet $\Sigma = \{a_0, a_1, \ldots, a_{l-1}\}$, initial state q_0, final state set F, transition function δ and blank symbol $\flat$. We describe a simulation of the computation of M on input $w \in \Sigma^*$. Let s a positive integer with $s \geq s_M(w)$ for $w \in L(M)$. By $s_M(w) \geq |w|$ also $s \geq |w|$. The parameter s determines the number of symbols and the length of strings needed for the simulation. It is, however, not important to know s in advance; instead, one can start with a small value of s, for instance $s = |w|$, and keep increasing it whenever needed – or one could guess s and simulate non-deterministically.

We consider the following sets of symbols, called epitopes and antibodies, respectively:

1. Epitopes $E = E_\Sigma \cup E_Q$ with

$$E_\Sigma = \{e_i^\Sigma \mid 1 \leq i \leq s\} \text{ and } E_Q = \{e_i^Q \mid 1 \leq i \leq s\}.$$

2. Antibodies $A = A_\Sigma \cup A_Q$ with

$$A_\Sigma = \{A_i^a \mid 1 \leq i \leq s, a \in \Sigma \cup \flat\} \text{ and } A_Q = \{A_i^q \mid 1 \leq i \leq s, q \in Q\}.$$

We assume that the antibodies A_i^q for $q \in F$ are marked so that their presence can be detected.

To encode the configurations of M we use a peptide with $2s$ epitopes as shown in Fig. 3. The antibodies will attach to their specific epitopes. A configuration $a_{i_1} a_{i_2} \ldots a_{i_{k-1}} q a_{i_k} a_{ik+1} \ldots a_{i_s}$ with $q \in Q$ and $a_{i_j} \in \Sigma \cup \{\flat\}$, for $j = 1, \ldots, s$, is encoded as

$$A_1^{a_{i_1}}[\ ]A_2^{a_{i_2}}[\ ]\cdots[\ ]A_{k-1}^{a_{i_{k-1}}}A_k^q A_k^{a_{i_k}}[\ ]A_{k+1}^{a_{i_{k+1}}}[\ ]\cdots[\ ]A_s^{a_{i_s}}$$

where [ ] denotes a free epitope, an epitope with no antibody attached. This situation is illustrated in Fig. 4

Figure 3: Peptide with state and symbol epitopes

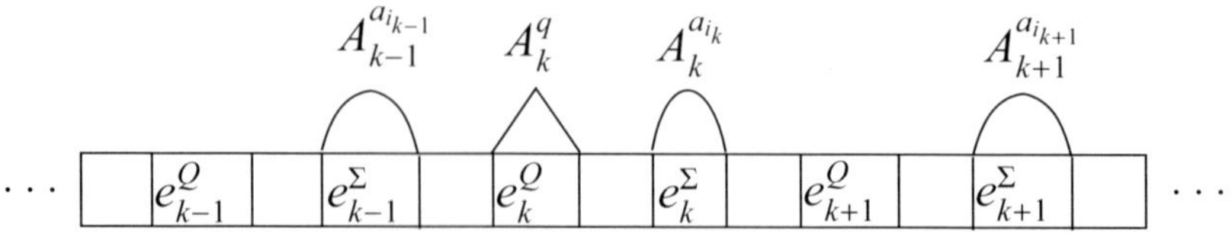

Figure 4: Peptide with antibodies attached to represent a configuration

Let $w = b_1 b_2 \cdots b_r$ where $b_j \in \Sigma$ for $j = 1, 2, \ldots, r$. The computation of M starts in state q_0 and reading the symbol b_1. We surround the initial configuration $q_0 b_1 b_2 \cdots b_r$ with enough copies of $\flat$ ($\flat$ denotes the blank symbol) so that the computation can be carried out within the space thus provided, when $w \in L(M)$. By assumption, $s-r$ copies of $\flat$ are needed and also sufficient; their correct distribution to the front and end is not important; as with the choice of the number s, also the distribution can be adjusted later or can be guessed non-deterministically. Thus, the configuration to be encoded by antibodies has the form

$$\flat\flat \ldots \flat q_0 b_1 b_2 \ldots b_r \flat\flat \ldots \flat$$

and consists of a total of s symbols in $\Sigma \cup \flat$. If t is the number of copies of $\flat$ preceding q_0, then the encoding is represented by

$$A_1^{\flat}[\ ]A_2^{\flat}[\ ]\cdots A_t^{\flat}A_{t+1}^{q_0}A_{t+1}^{b_1}[\ ]A_{t+2}^{b_2}[\ ]\cdots[\ ]A_{t+r}^{b_r}[\ ]A_{t+r+1}^{\flat}[\ ]A_{t+r+2}^{\flat}[\ ]\cdots[\ ]A_s^{\flat}.$$

This is the initial state of the simulation of M.

The simulation of a computation step involves the addition of appropriate free epitopes and antibodies depending on the present configuration and the transition function as follows.

Let $\ldots a_{i_{k-1}} q a_{i_k} a_{i_{k+1}} \ldots$ be the relevant part of the present configuration of M.

We distinguish the two cases of $\delta(q, a_{i_k}) = (q', a'_{i_k}, \mathrm{R})$ and $\delta(q, a_{i_k}) = (q', a'_{i_k}, \mathrm{L})$, right move and left move, respectively. In the former case, the next configuration is $\ldots a_{i_{k-1}} a'_{i_k} q'$ $a_{i_{k+1}} \ldots$ while in the latter case it is $\ldots q' a_{i_{k-1}} a'_{i_k} a_{i_{k+1}} \ldots$. We assume that enough symbols $\flat$ are present at the ends as required to perform the transitions. The peptide encoding of the present configuration is shown in Fig. 5.

To simulate the right move, we add an excess of free epitopes e_k^{Σ} and e_k^{Q}, and we then add the antibodies $A_k^{a'_{i_k}}$ and $A_{k+1}^{q'}$. The first step removes the antibodies $A_k^{a_{i_k}}$ and A_k^{q}. The second step attaches antibodies $A_k^{a'_{i_k}}$ and $A_{k+1}^{q'}$ which encode the new configuration of M. This encoding is shown in Fig. 6.

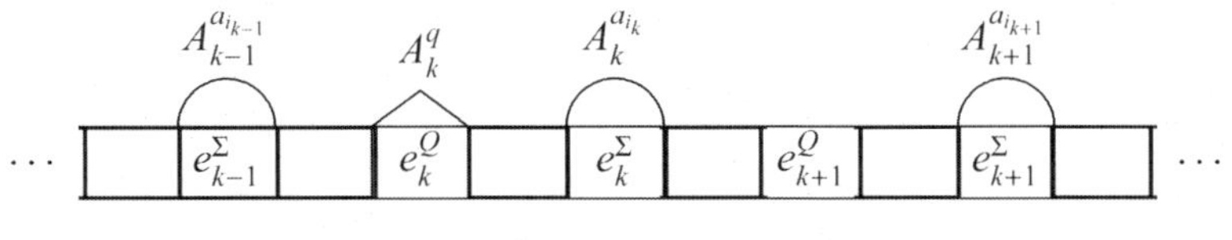

Figure 5: Peptide before applying the rule

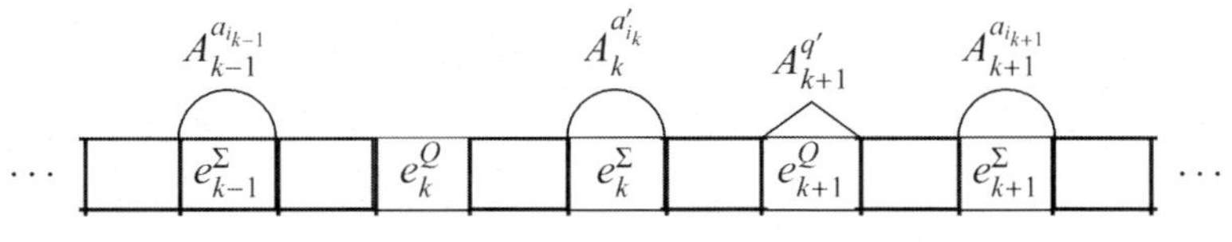

Figure 6: Peptide after applying the rule for a right move

The simulation of a left move is analogous. One adds an excess of free epitopes e_k^{Σ} and e_k^{Q}, and then the antibodies $A_k^{a'_{i_k}}$ and $A_{k-1}^{q'}$.

The end of the computation can be recognized by checking for the presence of antibodies A_i^q with $q \in F$ in the peptide sequence.

For each step of M, two steps of the peptide computation have to be performed, namely removal of antibodies and adding of antibodies. The length of the peptide needed for simulating the computation of M on input w is $O(s_M(w))$; exactly one peptide is required. The number of epitopes is $O(s_M(w))$ and that of antibodies is $O((m + l) \cdot s_M(w))$.

Finally, as mentioned above already, when s was chosen too small initially or when the distribution of the copies of $\flat$ was inadequate, one can extend the peptide sequence incrementally, until an appropriate situation has been obtained.

For each step of the Turing computation, the number of steps of the peptide computation is bounded. Hence the simulation runs in time $\Theta(t_M(w))$. □

The proof of Theorem 1 does not require that the Turing machine be deterministic; with minor modifications the simulation would also work for non-deterministic Turing machines with the same time and resource requirements.

3.2 Discussion

The simulation of Turing machines by peptide computing as presented in Section 3.1 is not totally convincing for two reasons:

First, the simulation is a "look-and-do" method, that is, it relies on an extraneous computing agent to interpret the computation of the Turing machine step-by-step and to simulate the observed behaviour by appropriate peptide-antibody interactions. This agent could, for instance, be a human taking notes of the steps of the Turing machine and selecting and mixing the required molecules. Thus, the simulation is not automatic.

Second, the simulation requires an unbounded number of epitopes and antibodies. The length of the peptide sequence in terms of epitopes and the number of antibodies needed are both approximately proportional to the amount of space used for the Turing computation.

To encode the antibodies and epitopes over a finite alphabet would increase the resource and time requirements. It seems that a comma-free encoding, even a solid code, would be needed to identify the locations of the epitopes in the peptide uniquely (see [28] for a survey of the relevant properties of codes). In this case, for s epitopes or antibodies to be encoded over an alphabet with r symbols, $r > 1$, about $c_1 \log_r s$ symbols are needed per epitope or antibody for some $c_1 > 1$. Thus, the size of the peptide is $O(s_M(w) \cdot \log s_M(w))$. Moreover, also the time bound increases by a factor of $\Theta(\log s_M(w))$ to account for the search for and the handling of, the appropriate epitopes and antibodies. While such an encoding is mathematically feasible, it is not clear at this point, whether this idea would work in the bio-chemical setting.

In the proof of Theorem 1, the transition function of the Turing machine is not encoded in the peptide system. To remedy this, we not only need an encoding but also a method for looking up instructions and for their execution.

In particular we need a method to encode the definition of the transitions in the interactions between peptides and antibodies. We also have to prepare the peptide sequences and antibodies in such a way that they select and execute the next transitions automatically. A concrete proposal for how to achieve this is presented in [7]; a full version of this work is in preparation [8].

4 Other Theoretical Models

4.1 Automaton Model

Several computing models have been defined which are inspired by peptide computing: binding-blocking automata (BBA) and their variants, like string binding-blocking automata (StrBBA) and rewriting binding-blocking automata (RBBA). A binding-blocking automaton is a finite automaton together with a binding and blocking function. These two functions help the machine to block some symbols on the tape from being read by the head. Hence the machine reads only the symbols which it needs and postpones the reading of other symbols to some other time. BBA have at least the power of a multi-head finite automata; StrBBA are even more powerful than BBA but not more powerful than random-context grammars; and RBBA are equivalent to Turning machines. It has to be noted that none of the above models are theoretical models for peptide computing; rather they are models inspired by the peptide computing model. For a study of these models see [6].

We list a few open questions regarding BBA and their variants:

1. In [6] it is shown that BBA accept languages which are not context-free. Do they accept all context-free languages? In other words, does the set of languages accepted by BBA contain the set of languages generated by context-free grammars, or are these sets incomparable?

2. What is the upper bound for the power of BBA with respect to the Chomsky hierarchy or other known hierarchies?

3. Can we obtain a necessary condition on the languages accepted by BBA and StrBBA similar to the pumping lemma?

In [5] a study of some complexity measures is presented. Assuming a complete characterization of BBA, we should be able to prove the infinite hierarchies presented in [5].

Although the heads of BBA do not move sequentially (sometimes they jump over some blocked symbols on the tape) BBA still do not capture the full parallelism of peptide computing. Hence it would be important to define an exact theoretical model for peptide computing. An attempt on these lines is made in [8].

4.2 Basic Operations

In [11], basic gate operations like *AND*, *OR* and *NOT* are defined using peptide-antibody interactions. This idea has been extended to model any Boolean circuit in [9]. The simple arithmetic operations of addition and subtraction of two binary numbers are modeled using peptide-antibody interactions in [10].

Although it is worth to see how to model the basic operations of a silicon computer, it will be more useful to define the basic operations which are special to the peptide computing model. This would help us to go a "molecular" rather than "silicon" way. When we use the silicon path we are locked into a special way of thinking and may not be able to make use of the immense parallelism inherent in molecular computing. Molecular operations, and this holds true for the binding of antibodies to peptide sequences, are not fast when compared to the operations of digital circuits. Hence when we try to solve a problem using peptide computing via gate operations we cannot gain much. So the main point here is to define the basic operations specifically for this model and then to formulate peptide algorithms accordingly. This approach would be similar to that of aqueous computing [25, 26].

5 Conclusions

In this paper we briefly described achievements in the new field of peptide computing. Many issues remain unresolved. We presented some open problems and proposals for their solutions.

In all the work so far in this area, one makes assumptions. Cross-reactivity or antibodies attaching to wrong epitopes are not considered. Also peptides are taken to be linear strands and antibodies attaching to some sites on those linear strands. In practice, the peptides fold and have three-dimensional structures, especially when they are sufficiently long. Moreover the antibodies attach to the peptides depending on folding. Hence one would have to develop a model including the three-dimensional structures and antibodies attaching to holes in such structures. This is a direction for further consideration.

We believe that there is a great potential in peptide computing, but the field is at an embryonic stage at present. Just defining the basic operations would not be enough. The problem of automating, the viability of bio-ware, error-control (like cross-reactivity) and implementation details for integrating the basic bio-ware into a general-purpose peptide computer have to be looked into.

References

[1] L. Adleman: Molecular computation of solutions to combinatorial problems. *Science* **266** (November 1994), 1021–1024.

[2] M. Aizawa, et al. (editors): *Bioelectronics, Research and Development Report 50.* CMC Press, 1984.

[3] M. Amos: *Theoretical and Experimental DNA Computation.* Springer-Verlag, Berlin, 2005.

[4] I. Antoniou, C. S. Calude, M. J. Dinneen (editors): *Unconventional Models of Computing, UMC'2K.* Springer-Verlag, London, 2000.

[5] M. S. Balan: Complexity issues in binding-blocking automata. In J. Dassow, M. Hoeberechts, H. Jürgensen, D. Wotschke (editors): *Pre-proceedings of International Workshop on Descriptional Complexity of Formal Systems.* 43–54, 2002. Revised version to appear in Journal of Automata, Languages and Combinatorics.

[6] M. S. Balan: *Computational Models using Peptide-Antibody Interactions.* PhD Thesis, Indian Institute of Technology, Madras, 2004.

[7] M. S. Balan, H. Jürgensen: On the universality of peptide computing. Technical report, Universität Potsdam, 2005.

[8] M. S. Balan, H. Jürgensen. Peptide computing: Universality and theoretical model. Manuscript, 2005.

[9] M. S. Balan, K. Krithivasan: Modeling boolean circuits using peptide-antibody interactions. In *Pre-proceedings of International Conference on Mathematical Biology*. 13–14, 2004. Proceedings in print.

[10] M. S. Balan, K. Krithivasan: Parallel computation of simple arithmetic using peptide-antibody interactions. *Biosystems* **76(1-3)** (2004), 303–307.

[11] M. S. Balan, K. Krithivasan: Realizing switching functions using peptide-antibody interactions. In N. Jonoska, G. Păun, G. Rozenberg (editors): *Aspects of Molecular Computing, LNCS 2950.* 353–360. Springer-Verlag, Berlin, 2004.

[12] M. S. Balan, K. Krithivasan, Y. Sivasubramanyam: Peptide computing: Universality and computing. In N. Jonoska, N. Seeman (editors): *Proceedings of Seventh International Conference on DNA based Computers, LNCS 2340.* 290–299. Springer-Verlag, Berlin, 2002.

[13] C. S. Calude, J. Casti, M. J. Dinneen (editors): *Unconventional Models of Computation.* Springer-Verlag, Singapore, 1998.

[14] C. S. Calude, M. J. Dinneen, M. J. P. Jimene, G. Păun, G. Rozenberg (editors): *Proceedings 4th International Conference Unconventional Computation, Lecture Notes in Computer Science*, 3699. Springer-Verlag, Heidelberg, 2005.

[15] C. S. Calude, M. J. Dinneen, F. Peper (editors): *Third International Conference, UMC 2002, Proceedings Lecture Notes in Computer Science*, 2509. Springer-Verlag, Heidelberg, 2002.

[16] C. S. Calude, G. Păun: *Computing with Cells and Atoms: An Introduction to Quantum, DNA and Membrane Computing.* Taylor and Francis Publishers, London, 2001.

[17] F. L. Carter (editor): *Molecular Electronic Devices.* Marcel Dekker, New York, 1982.

[18] F. L. Carter: The molecular device computer: Point of departure for large scale cellular automata. *Physics D* **10** (1984), 175–194.

[19] F. L. Carter (editor): *Molecular Electronic Devices II.* Marcel Dekker, New York, 1987.

[20] F. L. Carter, R. E. Siatkowski, H. Wohltjen (editors): *Molecular Electronic Devices.* North-Holland, Amsterdam, 1988.

[21] M. Daley, L. Kari: DNA computing: Models and implementations. *Comments on Theoretical Biology* **7**(3) (2002), 177–198.

[22] J. Dassow, H. Jürgensen: Soliton automata. *J. Comput. Syst. Sci.* **40** (1990), 158–181.

[23] R. Feynman: There's plenty of room at the bottom. In D. Gilbert (editor): *Miniaturization.* 282–296. Reinhold, New York, 1961.

[24] S. Hashida, K. Hashinaka, E. Ishikawa: Ultrasensitive enzyme immunoassay. *Biotechnol Annu Rev.* **1** (1995), 403–451.

[25] T. Head: Biomolecular realizations of a parallel architecture for solving combinatorial problems. *New Generation Computing* **19**(4) (2001), 301–312.

[26] T. Head, X. Chen, M. Yamamura, S. Gal: Aqueous computing: a survey with an invitation to participate. *Journal of Computer Science and Technology* **17**(6) (2002), 672–681.

[27] H. Hug, R. Schuler: Strategies for the development of a peptide computer. *Bioinformatics* **17** (2001), 364–368.

[28] H. Jürgensen, S. Konstantinidis: Codes. In G. Rozenberg, A. Salomaa (editors): *Handbook of Formal Languages*, 1. 511–607. Springer-Verlag, Berlin, 1997.

[29] L. Kari: DNA computing: tomorrow's reality. *Bulletin of the European Association for Theoretical Computer Science* **59** (1996), 256–266.

[30] N. Pisanti: DNA computing: a survey. *Bulletin of the European Association for Theoretical Computer Science* **64** (1998), 188–216.

[31] G. Păun, G. Rozenberg, A. Salomaa: *DNA Computing. New Computing Paradigms.* Springer-Verlag, Berlin, 1998.

[32] K. Shreder: Synthetic haptens as probes of antibody response and immunorecognition. *Methods* **20** (2000), 372–379.

M. Sakthi Balan[1], Helmut Jürgensen[1,2]
[1]Department of Computer Science,
The University of Western Ontario
London, Ontario, Canada, N6A 5B7
[2]Institut für Informatik, Universität Potsdam
August-Bebel-Str. 89, 14482 Potsdam, Germany

Kamala Krithivasan[3]
[3]Department of Computer Science and Engineering
Indian Institute of Technology, Madras,
Chennai - 600036, India

Formal Language Aspects of Natural Computing– 2006, pp. 77–84

An Introduction to Evolutionary Computation

Wim Hordijk

1 Introduction

Evolutionary Computation (EC) deals with computational methods that incorporate ideas and principles from natural evolution. More specifically, it includes a class of so-called *evolutionary algorithms* (EAs), which are heuristic search algorithms that can be used to search for good solutions to difficult problems. These algorithms try to "evolve" better and better solutions, as opposed to constructing one from scratch, and are particularly suitable for problems for which no known efficient (polynomial-time) algorithm exists, and for which exhaustive search is impractical. These evolutionary algorithms are fairly simple and general, and can be applied to a wide range of search and optimization problems. However, to get an appreciation for how and why they work, it is useful to understand the basics of genetics and natural evolution.

This paper is organized as follows. First, a brief overview of genetics and evolution is presented. Then, the concepts of search spaces, representations, and fitness functions are explained. Next, one particular evolutionary algorithm, namely the genetic algorithm, is explained in some detail, after which its advantages and disadvantages are discussed. A list of applications is then presented to give an idea of the kinds of problems that a genetic algorithm can be applied to successfully. Finally, some pointers to more information about genetic algorithms and evolutionary computation are provided.

2 Genetics and Evolution

In biology, a distinction is made between the genotype and the phenotype of an organism. The *genotype* is the genetic encoding of an organism, in particular the DNA molecules that reside in each and every cell in the body of an organism (whether it is a bacteria, fungus, plant, or animal). These DNA molecules contain *genes*, small subsets of DNA that encode for certain *traits* (or characteristics) of the organism, for example skin or fur color, number of legs, brain size, etc. More specifically, these genes, or pieces of DNA, are translated into proteins which form the basic building blocks of our bodies, and which perform the necessary functions for staying alive (metabolism, muscle activity, defense against disease, etc.). Together, the appearance of an organism and its functionality or behavior, form the *phenotype*, or the actual organism as it lives in the real world. So, in short, there is a one-way process where the genetic information (genotype) determines (together with environmental influences) the shape and functionality (phenotype) of an organism.

When two organisms of the same species get together to mate and create offspring, what happens at the genetic level is the following. Copies of the DNA molecules of both parents are mixed up, and the children receive part of their genotype from the father, and part from the mother. In other words, the genotype of an offspring individual is literally a (random) mix of the genotypes of its parents, a different mix for each child. This process is called *crossover*. Furthermore, during the process of copying the genetic information from parents to offspring, sometimes small errors occur, called *mutations*. These happen only very infrequently, but they do introduce some (possibly) new variation in the genetic information. As a result of these crossover and mutation processes, the offspring individuals (i.e., the next generation) will have some variation compared to their parents. Some traits of an individual in the next generation will have come from its father, and some from its mother. Moreover, occasional mutations might be expressed in some entirely new variation in a certain trait. In short, there is descent with variation, where the variation is introduced at the level of the genotypes (through crossover and mutation), and expressed at the level of the phenotypes.

In the real world, not all individuals that are being born will survive and create offspring themselves. For example, in a rabbit population, most baby rabbits will not survive until adulthood: they might get eaten by predators, they will have to compete with other rabbits for food sources, or they may encounter any other natural or accidental causes of death before they have a chance to mate. Consequently, only a small portion of each generation of rabbits will actually create offspring themselves, and pass on their genetic information to the next generation. In particular, those rabbits with "good" traits, such as those able to run fast or with a fur color that allows them to hide in the bushes to escape predators, will (on average) survive and create offspring. Individuals with "bad" traits (or "not-so-good" traits), however, will not be very successful (again on average) in surviving, and will not be able to create offspring (or at least not as many as the more successful individuals). This process, summarized by the familiar expression "survival of the fittest", is called *natural selection*: advantageous traits (or variations in traits) will on average be preserved and proliferate in the population, and disadvantageous traits (or variations) will on average disappear again, because they are not passed on to subsequent generations. Over time, the population as a whole will thus slowly change, or evolve, due to genetic variation, and adapt to cope better with the environmental challenges ("survival") through natural selection.

This, in a nutshell, is how evolution and natural selection happen in the real world. Of course this is a somewhat simplified picture, but these main principles are what is of importance here. One particular view of evolution that has become very popular is that of evolution as a *problem solver*. The "problem" is that of survival and reproduction, and evolution through natural selection is the mechanism of "solving" the problem. In other words, evolution searches through the space of possible DNA molecules (genotypes) for those that encode for successful phenotypes that are able to survive and create offspring. Those genotypes that encode for successful (fit) phenotypes will, on average, be used more often than less fit ones to produce new genotypes (through genetic variation) to make up the next generation. This way, over subsequent generations, the individuals in the population will become more and more successful at "solving the problem."

3 Search Spaces, Representations, and Fitness Functions

To use the principle of evolution in computer search algorithms, first we need to review the concepts of search spaces, (genetic) representations, and fitness functions.

3.1 Search spaces

The *search space* of a given problem is simply the space of all possible (candidate) solutions to that problem. For example, consider the subset sum problem: given a set of N numbers, is there a subset S of these numbers such that the sum of the numbers in this subset add up to some given number M? Or in the optimization version of this problem one could ask for a subset for which the sum is as close as possible to M. Obviously, the collection of all candidate solutions to this problem (the search space) is the set of all possible subsets of N numbers, of which there are 2^N (each number is either in a subset, or it is not). As another example, consider the traveling salesman problem (TSP). Here, the object is to find the shortest tour along N cities (which are linked by roads with certain distances), such that each city is visited exactly once. Here, the search space is the collection of all possible tours along these N cities, of which there are $N!$.

3.2 Representations

Sometimes it is convenient to represent candidate solutions to some problem in a simple way, e.g. with strings of characters, which can be easily manipulated using a computer program. For example, in the subset sum problem each candidate solution can be represented by a bit string (a string of 0s and 1s) of length N. Given a particular bit string, the corresponding candidate solution can be easily reconstructed by including the number x_i in the subset if its corresponding bit value (i.e., the ith bit in the string) is 1. Similarly, possible tours along N cities can be represented by permutations of the numbers 1 to N. The order of the numbers in a given permutation simply indicates the order in which to visit the cities in the corresponding candidate solution, or tour. This way of representing candidate solutions to a given problem is analogous to the genotype-phenotype distinction in biology. The representations (bit strings, permutations) are the genotypes and the corresponding candidate solutions (subsets, tours) are the phenotypes. Thus, the search space of subsets is represented by the space of bit strings (of length N), and the search space of tours is represented by the space of permutations (of length N).

3.3 Fitness functions

In order to apply an evolutionary algorithm to a given problem, we need some way of evaluating candidate solutions. This is done with a fitness function. A *fitness function* is a function that takes as input a representation (or genotype), translates this into the corresponding candidate solution (or phenotype), tests this candidate solution on the given problem, and then returns a number that indicates how good this solution is, i.e., its *fitness*. For example, in the subset sum problem, the fitness function takes as input a bit string, from this it constructs the corresponding subset of numbers, and compares the sum of this subset to the given number M. In the TSP problem, the input is a permutation, which gets translated into the corresponding tour, and the length of this tour is then calculated.

Since we generally view evolution as a maximization problem, i.e., fitness should be maximized, a fitness function is usually constructed in such a way that a higher fitness value is assigned to better solutions. In the examples given, we are dealing with minimization problems (as is often the case in real-world problems), so better solutions will have lower values (such as a smaller difference with the given number M, or shorter tour lengths). However, every minimization problem can be easily converted into a maximization problem, and the fitness function could, for example, return the inverse of the calculated fitness values, or some large number minus the calculated value. So, here we will only consider evolutionary algorithms as maximizing the fitness values, in analogy with biology. The fitness function for a given problem (whether it is a minimization or maximization problem) can always be written in the appropriate form.

4 Genetic Algorithms

As one particular example of an evolutionary algorithm, we will now give an overview of the *genetic algorithm* (GA). GAs were originally developed by John Holland in the 60s and 70s [4]. For a detailed introduction, including examples and applications, see [5, 3, 6]. A GA searches for good solutions to a problem by maintaining a population of candidate solutions (suitably encoded), and creating subsequent generations by selecting (on average) the current best solutions and using operators like crossover and mutation to create new candidate solutions. Thus, better and better solutions are "evolved" over time.

To apply a GA, first a suitable encoding (a set of genotypes) needs to be chosen to represent candidate solutions to the given problem. Examples of some possible representations were given in the previous section. Next, an appropriate fitness function needs to be constructed. As described above, this function is used to translate a genotype to its corresponding candidate solution, and then evaluate this solution on the given problem. A "better" solution will have a higher fitness value. The choice of representation and the construction of the fitness function depend, of course, largely on the problem at hand. Given a representation and fitness function, the algorithm now roughly works as follows (implementation details are discussed below):

Genetic Algorithm

1. **Initialization:** Initialize a population of P individuals (genotypes) at random.
2. **Fitness assignment:** Evaluate the individuals in the current population (calculate their fitness values) using the fitness function.
3. **Selection:** Create a (temporary) mating pool of "parents" by selecting (with replacement) P individuals from the current population, placing a copy of each selected individual in the mating pool. The probability of an individual being selected is proportional to its fitness relative to the rest of the population (highly fit ones might get selected more than once, less fit ones less often or not at all).
4. **Reproduction:** While there are still parents in the mating pool, do:
    (a) Take the next two parents from the mating pool.
    (b) Perform crossover on these two parents, creating two "children".
    (c) With a small probability, apply mutation to the children.
    (d) Place the children in the new population.
5. **Replacement:** Replace the old population with the new population.
6. **Repeat:** Go back to step 2, unless some stopping criterion is reached (e.g., a certain number of generations or a certain level of fitness).
7. **Stop:** Return the best individual in the final population.

During the initialization step, individuals are created completely at random. For example, when the chosen representation is bit strings, P (the population size) random bit strings are constructed by simply choosing for each bit position in each individual either a 0 or a 1 with equal probability. Random permutations can be created by filling an array with N random numbers chosen uniformly between 0 and 1, and assigning values from 1 to N to each position in the array according to the rank of its random number. In other words, the position with the largest (random) number gets value 1, the second largest number gets 2, and so on, until the position with the smallest number, which gets value N.

Next, the individuals in the population are evaluated, and their fitness values are calculated. Based on these fitness values, individuals are now selected to contribute their genetic information to the next generation. The idea is that individuals with a high fitness, compared to the population as a whole, will have a high chance of being selected, and individuals with a low fitness value will have a low chance. One way of implementing this is using *roulette wheel selection*. In this selection method, each individual will have a chance $f_i / \sum_i f_i$ of being selected, where f_i is the fitness of individual i. One can imagine this as dividing a roulette wheel into P slices, where the size of each slice i is proportional to the fitness of individual i relative to the population total. Now a roulette ball is rolled P times, and each time the individual corresponding to the slice in which the ball ends up is copied to the (temporary) mating pool. Obviously, individuals with larger slices (high fitness values), have a higher chance of having the ball end up in their slice, i.e., of being selected. Consequently, individuals with above average fitness values (compared to the current population average) will most likely end up with multiple copies in the mating pool, and individuals with below average fitness values only with a few or none, in proportion (statistically) to their fitness values relative to the population total. So, selection, or "survival of the fittest", is implemented based on how "good" (fit) the current solutions in the population are as compared to each other. Of course there are other ways of implementing the selection method, such as tournament selection, where each time two individuals are picked from the population (with equal probabilities), and the most fit of the two is copied into the mating pool. This is then repeated P times. Still other methods exist and have been used as well, but the main idea is that individuals are selected based on their fitness values.

Once the mating pool has been created, the individuals in this pool are allowed to reproduce and create offspring. As long as there are still individuals in the mating pool, the next pair of individuals, or "parents", is removed from the pool, and the crossover and mutation operators are performed on them, creating two "children". These children are then placed in a new population. For bit string representations, the crossover operator works as follows. Consider two parents as in the example below (the all 0s and the all 1s strings are chosen for clarity of the example). First, a crossover point (the vertical bar) is randomly chosen somewhere between the first and last bit positions, in this case at the 4th position. Next, the first part of the first parent (up to the crossover point) is combined with the last part of the second parent (after the crossover point) to create the first child, and vice versa for the second child.

```
0000|000000   -->   0000|111111
1111|111111   -->   1111|000000
```

Often, crossover is only applied with a certain probability p_c, which is usually set in the range [0.6;1.0]. In this case, for each pair of parents, a random number between 0 and 1 is drawn first. If this number is smaller than the crossover probability, then a random crossover point is chosen and crossover is applied as explained. Otherwise, the two children will be identical to the parents (i.e., no crossover is applied). Of course there are many variations on this basic crossover operator (known as one-point crossover). We can also use two or more crossover points, or even decide (randomly) for each individual bit in the children from which parent it will come. And, obviously, for different kinds of representations (such as permutations), different crossover operators are required to ensure that the resulting children are still valid representations for candidate solutions (for example, applying one-point crossover to two permutations does generally not result in two new valid permutations, but more suitable crossover methods are available for permutations as well).

Next, with some probability p_m mutation is applied to the children, where a bit is flipped (i.e., a 0 changed into a 1 or vice versa). However, this is usually only done with a very low probability, so that (on average) only a few children actually undergo a mutation in each generation. This operator is mostly performed to prevent the population from converging too quickly (i.e., to maintain some diversity), and also to introduce some new variation, if possible. For example, if all the individuals in the population have a 1 at a certain bit position, then by using crossover alone it is impossible to get a 0 at that position. However, with mutation, such a change is possible. Here, too, different representations will require different types of mutation operators.

This reproduction process (crossover and mutation) is then repeated for the next two individuals (parents) in the mating pool, and then the next pair, and so on, until P new individuals (children) have been created. This new population then replaces the previous one, thus forming the next generation. Now, the fitness values of the new individuals can be calculated again, and the process of selection and reproduction is repeated to create subsequent generations. This way, the algorithm searches through the space of candidate solutions (by "manipulating" genotypes), trying to create better and better ones based on the current best individuals in the population and recombining (mixing) their genetic information. The algorithm usually stops when a given number of generations or some time limit is reached, or when no significant improvements are found anymore after some amount of time or number of generations.

In short, GAs are "just another" class of heuristic search algorithms that can be applied to problems for which there are no efficient algorithms available, and for which the search space is too large to perform an exhaustive search. However, in practice it turns out that they perform quite well, often finding better solutions than other search algorithms (such as, for example, simulated annealing or hill climbing) or greedy methods. Their success is often ascribed to the fact that they use a population of candidate solutions (instead of looking at one candidate solution at a time), and that they combine global search (crossover) with local search (mutation). Thus, GAs provide a class of widely applicable, yet powerful search algorithms that are inspired by natural evolution, and form an important part of what is known more generally as evolutionary computation.

5 Advantages and Disadvantages of GAs

Even though GAs seem to work quite well in practice, there are several disadvantages attached to them as well. In particular, since they are fairly general search algorithms, it can be difficult to "fine tune" the algorithm to work well on any specific problem. The more complicated a problem is, the more work needs to be done to make the GA perform well. As already mentioned earlier, the choice of representation and the construction of the fitness function depend highly on the problem at hand. A bad choice for any of these can make the algorithm perform poorly. Unfortunately, it is still not known very well how to make the best choices, or how to construct the best fitness function given a problem. Most of this relies on trial and error, and often much thinking and testing needs to be done before the algorithm performs reasonably well. Furthermore, there are many different operators to choose from (different selection, crossover, and mutation methods), and several parameters to set (population size, crossover and mutation rates, etc.). Again, it is not always obvious what the best choices are, and here, too, trial and error are often the only option. Finally, there is no convergence guarantee. Whatever the best solution is that the GA finds, it is generally unknown how close this "best" solution is to the global optimum (if we knew what the global optimum is, we would not need a GA!). So, it is advisable to run the GA several times, each time starting with a different (random) initial population, and perhaps using different parameter settings. The best solution over all these runs is then taken as "the solution". If this is better than all other solutions known so far, then at least we have achieved something...

Below is a brief summary of the advantages and disadvantages of GAs:

Advantages

- GAs are fairly easy to understand and implement.
- GAs are applicable to a wide range of problems.
- GAs are easy to parallelize (e.g., simultaneous calculation of fitness values).

Disadvantages

- Choosing a good representation and constructing a good fitness function depend on the problem at hand, and can be difficult.
- There are many operators (selection, crossover, mutation) to choose from, and several parameters to set, and it is often not clear what the best choices are.
- There is no convergence guarantee. Just run the GA many times and hope for the best!

6 Applications

To give an idea of the kinds of problems that GAs have been applied to successfully, here is an overview of some applications (not meant to be exhaustive though!):

- Aircraft wing design (optimizing parameters that determine the shape of a wing)
- Optimal pipeline flow (optimizing, e.g., the flow of oil through a network of pipelines to satisfy customer demand)
- Scheduling problems (job-shop scheduling, etc.)
- Routing problems (TSP, network routing)
- VLSI technology, chip design (minimizing connections and communications)
- Telecommunications networks (optimal location of access points, etc.)
- Coding methods for data transmission (optimizing coding methods and their parameters)
- Robotics (evolution of robot controllers)
- Stock market prediction (making money with money... and evolution!)

And, of course, there are many, many more applications. It is impossible to list all existing GA applications here. However, the next section gives some pointers to where more information on genetic algorithms and evolutionary computation can be found.

7 More Information

The original book on genetic algorithms by Holland [4] is fairly technical and mathematical, and is perhaps not the best introduction to GAs. However, his later article in *Scientific American* [5] is highly recommended, as it is written for a general audience and introduces GAs at a high level that is very readable. Furthermore, two standard textbooks with good introductions but also with much detail and many examples are [3, 6]. In [2], a good overview of the different GA methods (such as selection, crossover, and mutation) and their implementations can be found. As mentioned above, it is impossible to list all the different problems that GAs have been applied to. However, the various proceedings of the international conferences on genetic algorithms provide one of the most complete collections of GA applications. And of course the internet is a rich source of information. One good starting point is the GA Archives, maintained by the US navy: `www.aic.nrl.navy.mil/galist`.

Information on evolutionary computation in general is also widespread. An excellent source is the *Evolutionary Computation* journal, published by the MIT Press, or equivalently, the *IEEE Transactions on Evolutionary Computation*. Another useful source is [1], and many application areas can be found in the proceedings of the international conferences on evolutionary computation. On the internet, the *Hitch-Hiker's Guide to Evolutionary Computation* is a wonderful starting point, with many links to other pages: `www.cs.bham.ac.uk/Mirrors/ftp.de.uu.net/EC/clife/www`.

References

[1] T. Bäck, D. B. Fogel, and Z. Michalewicz (editors). *Handbook of Evolutionary Computation*. Oxford University Press, 1997.

[2] L. Davis (editor). *Handbook of Genetic Algorithms*. Van Nostrand-Reinhold, 1991.

[3] D. E. Goldberg. *Genetic Algorithms in Search, Optimization, and Machine Learning*. Addison-Wesley, 1989.

[4] J. H. Holland. *Adaptation in Natural and Artificial Systems*. University of Michigan Press, 1975.

[5] J. H. Holland. Genetic algorithms. *Scientific American*, 267(1):66–72, 1992.

[6] M. Mitchell. *Introduction to Genetic Algorithms*. MIT Press, 1996.

Wim Hordijk
Karunya Institute of Technology and Sciences
Deemed University
Karunya Nagar, Coimbatore - 641 114
email: wim@santafe.edu

Formal Language Aspects of Natural Computing– 2006, pp. 85–101

From Molecular Computing to Modelling with Conformons and Computing by Observation

Pierluigi Frisco

Abstract

A short overview on Molecular Computing, with a particular focus on Membrane Computing, is presented.

The modelling potential of (biological) processes via Conformon-P systems and the computational paradigm of Computing by Observing are outlined.

This article does not use a technical language and it is targeted to young researchers. Some suggestions for research are given.

1 A dialog

One day a biologist (**B**) was explaining some cell biology to a computer scientist (**CS**). In order to make his presentation more understandable he brought with him some video animations sketching the activity of a (eukaryotic) cell.

B: This animation shows the activity of cutting machineries, called *restrictions enzymes*, on DNA molecules. You can imagine a DNA molecule just as a piece of text, and a restriction enzyme as a small machinery able to recognise a specific part of the text and to break it. For instance we can imagine a restriction enzyme recognising the word 'the' and breaking the text after the 'the'.

CS: I understand. The restriction enzyme and the DNA molecule just wander around in the cell and when they meet the DNA molecule can be cut by this machinery.

B: Exactly!

CS: But I also see in this movie that some DNA molecules become longer.

B: Yes. Two DNA molecules can be joined together by a gluing machinery called *ligase*.

CS: So, in a cell, DNA molecules can be cut and glued together.

B: More or less. This affirmation is true for what it concerns this explanation. This cutting and pasting process is called *splicing*.

CS: But, can any two DNA molecules be glued together even if they where not linked before?

B: Well, there are some conditions, but for sure this can happen. New 'texts', that is, new DNA molecules, can be formed.

CS: I also see all these things in a cell moving around, passing in and out these rooms.

B: What you called 'room' is a *membrane compartment*. It can be indeed seen as a room inside a cell. Cells are full of this rooms, each of them has a different function. Some of them can allow only some things, not all, to enter them.

CS: So, for instance, it can be that in a room some molecules are processed by some machinery and only the processed molecules can pass to another room, there being further processed, maybe in combination with molecules coming from another rooms, and so on.

B: Surely this happens in a cell.

CS: This activity looks to me very much alike to a computation performed with molecules. In a very basic way I can tell you that computation starts from the input, for instance the original DNA molecules, some operations, the machinery cutting and pasting the DNA molecules, and the output, the DNA molecules that, for instance are present in a specific room of a cell.

B: Can you make an example, how can this be seen as computation?

CS: Well, if for instance there are many DNA molecules of the form: '<Len-<Tom-<Erik-' and 'runs>sleeps>eats>' (the input), one machine that cuts after a '-', another that cuts after a '>', then it is possible to have '<Len-', '<Tom-', '<Erik-', 'runs>', 'sleeps>', and 'eats>'. The gluing machinery can glue for instance '<Len-' and 'runs>' so to form '<Len-runs>' (the output). If we imagine that there is a room that can let in only DNA molecules with one '<' and one '>', then the cell computes the actions that 'Len' can take.

B: I understand. But even if there are many, do not believe that there are machines cutting all what we want and membranes letting in all what we want to go in. In between others, biologists are working on discovering new cutting machinery and maybe also to create cutting machinery that cut where we want.

CS: Well, it is a pity that we cannot do all what we want. Anyhow, this way to compute, cutting and pasting things and having rooms that let only specific things to get in, is new to me and I would like to investigate what kind of problems can be solved in this way.

B: What do you mean with "this kind of problems"? - smiling - To 'compute' how many things Len can make?

CS: Of course not! In Computer Science there are many categories of problems, more difficult, less difficult, ... It would be nice to study which problems a system with cutting and paste, that I can call *Splicing system*, and another with membranes, that I can call *Membrane system*, can solve. This kind of way to compute inspired by how molecules interact could be called Molecular Computing. This study could lead to deepen our understanding of what computation is and, ultimately, to the creation of machines computing with molecules.

B: Of course all this is 'in theory', not in a biological laboratory.

CS: Yes, for sure it would be a pen and paper work, but probably something could also be implemented in a lab.

2 Introduction

The fantastical dialog presented in the previous section wants to underlying the basic idea that drives *Molecular Computing*: the study of processes present in nature as computational processes.

This study opens a broad range of investigations:

- ⋆ is it possible to use molecular processes to implement computation?
- ⋆ what are the differences, from a computational point of view, of several molecular processes?
- ⋆ can we look to other biological (not only molecular) processes as computational processes?
- ⋆ what can we learn from the modelling of biological processes using computational devices?
- ⋆ ...

The potential of this research is incredible, it could:

- ⋆ bring to light new computational devices (using, for instance, a new kind of hardware);
- ⋆ develop new computing paradigms (think, for instance, to Evolutionary Algorithms and Neural Network);
- ⋆ let us devise new ways to manipulate/define information (think, for instance, to recursion);
- ⋆ lead us to a better understanding of biological processes;
- ⋆ ...

In Section 3 we give a short overview on Molecular Computing starting from its origins until its latest developments. In Section 4 we give a short introduction to Membrane Computing and we describe conformon-P systems in particular. In Section 5 we shortly outline the idea on which computing by observation is based.

3 A short overview on Molecular Computing

The idea to perform computation on a molecular scale it is not new. For example, already in 1973 C. Bennett discusses in [10] how RNA could be used as a physical medium for implementing (reversible) computation, linking in this way the functioning of RNA polymerase to a Turing machine. In [11] the same author gives a description of a hypothetical computer using a macromolecule similar to RNA to store information, and imaginary enzymes to catalyse reactions on the macromolecule.

However, it was not until 1994 that a molecular implementation of a computation took place. In his seminal paper [1] L. Adleman describes the solution of an instance of the directed Hamiltonian path problem in a laboratory experiment using DNA molecules and molecular operations on them. This experiment was striking for its simplicity: the operations used by Adleman are the standard biomolecular operations used in biological laboratories.

The Adleman's experiment had tremendous impact: it has stimulated a whole avalanche of research dealing with both the experimental and theoretical aspects of computing using molecules. A new and exciting research area, *Molecular Computing*, was born.

The potential of using DNA molecules for solving hard computational problems has been investigated from both experimental and theoretical points of view. For example, theoretical solutions to the satisfiability problem (SAT) are discussed in [63, 49, 66, 36], while experimental solutions of small instances of SAT are described in [90, 13]; a theoretical solution to 3-colorability problem is presented in [2]; an experimental solution of a small instance of the maximal clique problem is reported in [76]; a theoretical solution for circuit satisfiability problem is given in [12].

An important research stream of Molecular Computing is molecular implementation of various models of computation. Thus, for example, molecular implementation of Turing machines was considered in [7, 8, 10], and also in [88] where molecular encodings of the transition table of a Turing machine are given in great detail. E. Winfree pioneered a thorough investigation of implementation of standard models of computation based on self-assembly, see [100]. Among others he considered right-linear grammars, context-free grammars, and cellular automata [101, 89, 99, 103, 104, 91, 92, 5]. An interesting autonomous implementation of finite automata using restriction enzymes has been presented in [9], on the same line also [93, 15].

One can distinguish two (not necessary disjoint) research trends of Molecular Computing: Molecular Computing *in vitro* and Molecular Computing *in vivo*. The part of Molecular Computing discussed above concerns mostly Molecular Computing *in vitro*. Research in Molecular Computing *in vivo* can be again divided in two streams: implementing computational devices in living cells, and discovering/analysing computational processes taking place in living cells. An example of the former line of research is, e.g., the implementation of boolean circuits [98, 97] or finite state automata [74] in living cells, while an example of the latter line of research is the investigation of gene assembly in ciliates [52, 61, 24, 22, 23].

A very interesting research concerns a combination of various types of natural computing; for example combining evolutionary algorithms [71] and Molecular Computing [62]. The advantage of such a combination is the fact that DNA molecules form a very good medium to store information that can be modified in its evolution by many types of factors such as variable-rate mutation, rearrangement by transposition, retroviral insertion and editing [21, 26, 73, 105].

Research in Molecular Computing has resulted in a big number of models with some of them being only theoretical and others that gave also rise to experimental implementation. It is beyond the scope of this paper to list all the models, however we would like to mention some of them: *aqueous computing*, whose essence is the writing on DNA molecules dissolved in water, [42, 43, 44]; *microflow reactors*, in which streams of DNA molecules pass through tiny microflow channels where some general programmable operations can be performed, [69, 70, 94]; *sticker systems*, which model the self-assembly of DNA molecules, [53, 32, 82, 95]; *Watson-Crick automata*, in which tapes are double stranded sequences of symbols related by a complementarity relation, [31, 68]; *insdel systems*, which are based on context sensitive insertion and deletion of symbols/strings, [51, 54]; *forbidding-enforcing systems*, which abstract the effects of chemical reactions, [25, 95]; *membrane computing* Section 4.

One should also mention research concerning storage potential of biomolecules such as DNA. This potential is really enormous: one gram of dry DNA occupies a volume of approximately one cubic centimeter and it can store as much information as approximately one trillion CD's, this means 1 bit per cubic nanometer; moreover, a dry DNA molecule has a stability of about one billion years. Examples of papers dealing with DNA-based-memory are [6, 86, 55, 18, 87]. A broad overview on the field of DNA computation can be found in [4].

To summarise: Molecular Computing is a fast growing interdisciplinary research area involving (among others) computer science, molecular biology, chemistry, mathematics, and physics, and establishing bridges between information processing, nanotechnology and biology. Most importantly, it broadens and deepens our understanding of the theory and practice of computing.

4 Membrane computing and modelling

In this section we will give a short overview on *Membrane Computing*, describing its motivations and trends. Then we will focus on *conformon-P systems*, an abstract computability model that fits well the modelling of biological processes. We will informally define these systems and give some suggestions for research both in the area of theoretical computer science and in the one in modelling.

4.1 Membranes in a cell

The word *cell* was coined in 1665 by Robert Hook who, looking at thin slices of cork through a microscope, observed tiny, hollow, room-like structures which reminded him of rooms that monks lived in. Actually what Hook was observing were what we call now *cell walls* (because the cork cells were not alive), i.e., *membranes* surrounding a cell.

Cells have several membranes - each of them is a thin and pliable layer. All membranes present in a cell are based on *lipid molecules*. Lipid molecules in cell membranes have a hydrophilic ('water-loving') or *polar* end and a hydrophobic ('water-fearing') or *nonpolar* end; Figure 1 depicts a phospholipid, i.e., a lipid having a polar head and two hydrophobic tails.

Lipid molecules can spontaneously assemble to form *bilayers* - Figure 2 illustrates a section of a bilayer composed by phospholipids.

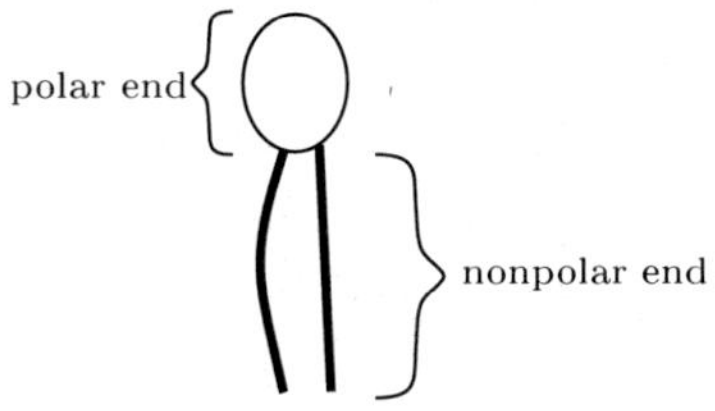

Figure 1: A phospholipid

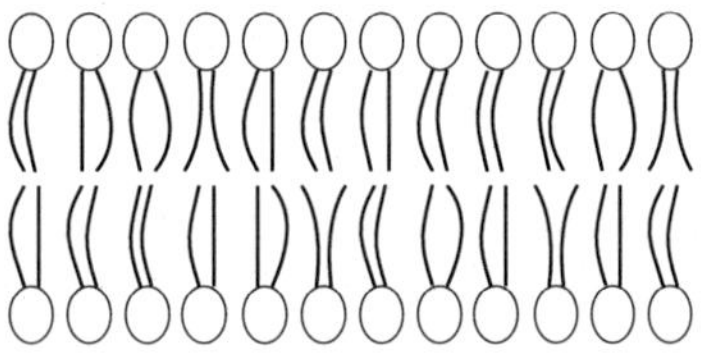

Figure 2: A bilayer of phospholipids

Among many membranes present in an eukaryotic cell we can distinguish the *plasma membrane* which delimits the cell from the environment and the *nuclear envelope*, composed by the *inner nuclear membrane* and the *outer nuclear membrane*, which encloses the DNA.

One of the main functions of membranes is the formation of compartments (enclosed by membranes) where specific molecular processes take place. Another important function of membranes in cells is the transport and communication (within a cell and between cells). The combination of these functionalities allows a cell, or a component of it, to keep certain chemicals inside and making sure that other chemicals stay outside.

4.2 Membranes in computing

The subdivision of a cell into compartments delimited by membranes has been of inspiration to G. Păun for the definition of a class of (distributed and parallel) model of computation called *membrane systems* [79]. In the seminal paper the author considers systems based on a hierarchically arranged, finite *cell-structure* consisting of several cell-membranes embedded in a main membrane called the *skin*. The membranes delimit *regions* where *objects*, elements of a finite set, and evolution rules can be placed - this is illustrated in Figure 3.

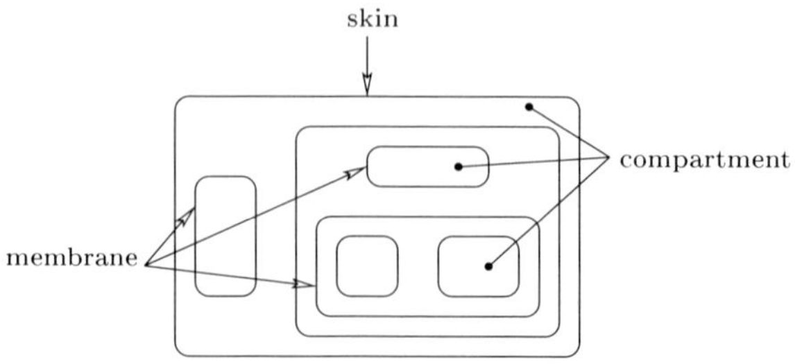

Figure 3: A schematic cell-structure

Originally, three types of membrane systems were considered: transition, rewriting and splicing (this classification depends on the kind of objects and the way they are processed). Objects evolve according to evolution rules associated with compartments and they can pass through membranes to adjacent compartments. A computation starts from an initial configuration of the system, defined by a cell-structure with objects and evolution rules in each compartment, and terminates when no further rule can be applied.

Research on membrane systems, also called ‘P systems’ where ‘P’ stands for ‘Păun’, has really flourished [81, 106]. Many new features as well as variants of original features were considered in the literature - we give now some examples. The membrane structure can be extended to a general graph [84, 85]. The membranes themselves can be actively involved in the computation allowing the model to dissolve, create or duplicate them [80, 41]. Different kinds of objects can be considered: without any structure such as elements of multisets, with a well defined structure such as strings, or something in-between such as conformons [67, 78, 33]. General features related to rules such as the presence or absence of maximal parallelism can be considered [58, 33]. Also the generated language can be defined in many ways: as the (multi)set of objects present in a specific compartment when a computation terminates, or the sequence of symbols sent into a compartment, or the trace of a specific object [20, 30, 37].

One can distinguish three main lines of research concerning membrane systems:

1. establishing the generative power (many class of P systems were proved to be computationally universal, i.e., as powerful as a Turing machine);

2. using them to develop algorithms for solving computationally hard problems (NP problems were solved in polynomial time [59, 75, 65]);

3. using them as modelling platform [19].

The algorithms related to item 2 in the previous list exploit the inherent parallelism of P systems and the capacity for dynamic changes in the structure of the system while it operates. This dynamic change allows, for example, the increase in the system's workspace while in operation and, in turn, reflects the original inspiration in the dynamics of cell division and self-organised compartmentalization.

This capacity of membrane systems is also unique so far in the realm of abstract computational architectures. This has led to the development of new complexity classes (LMC, MC Chapter 7 in [81]). These classes, by embracing the concepts of both inherent parallelism and workspace division/replication are the first, therefore, to relate directly to the real computational capacities of living systems. The standard classification [77] (P, NP, PSPACE, NPSPACE, etc.) is based only on sequential computing models, while the NC class for parallel systems assumes a predetermined and fixed number of processors.

Suggestion for research: As indicated in [81] these new complexity classes need further investigation.

What happens if we allow membrane division/creation but not maximal parallelism? (this is a much more realistic scenario as maximal parallelism is not present in cells)

4.3 Conformon-P systems

In [38] Frisco & Ji introduced a variant of membrane systems called *conformon-P systems*. This variant, later studied also in [33, 34, 35, 39], is based on simple and basic concepts inspired by a theoretical model of the living cell centred around *conformon* [46, 48].

The concept of conformon was introduced in molecular biology independently in [40] and [96]. The common part of the two definitions is the conformational deformation of (macro) molecules in a cell. In [40] attention is focused on the free-energy associated with the localised conformational strain: a conformon is seen as a sequence specific mechanical strain that provides both free-energy and Shannon information.

Following this, conformons and conformon-like entities have been classified into 10 families based on their biological functions [47]. The term was adopted in [56] and [57], where the authors developed the first quantum mechanical representation of the conformon concept.

A conformon-P system has conformons, a name-value pair, as objects; they can *interact* (exchanging part of their value) according to evolution rules present in membrane compartments. Moreover conformons can *pass* from one membrane compartment to another depending on their value. A simple conformon-P system is illustrated in Figure 4.

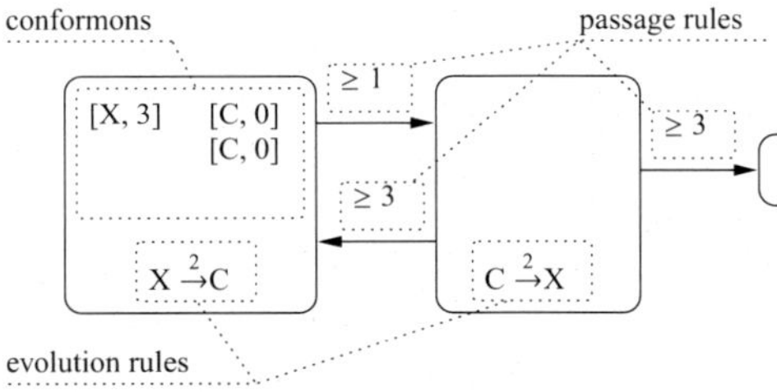

Figure 4: A conformon-P system

Some computational aspects of conformon-P systems have been studied to date [33, 38, 39].

These investigations led to the characterisation of an initial hierarchy composed by: *basic conformon-P system*, *proper conformon-P system*, *conformon-P system with priorities* and *conformon-P system with unbounded value*.

Variations on the structural and related constraints of conformon-P systems have clear influences on their computational power (i.e., the kind of problems that can be solved). For example, it has been proved that the computational power of conformon-P systems with priorities and conformon-P systems with unbounded value is equivalent to that of so-called program machines [72], that the power of conformon-P systems is equivalent to that of partially blind program machines, and that basic conformon-P systems can only generate finite languages.

Suggestion for research: It would be interesting to continue to modify the definition of conformon-P systems given in [33], removing or adding features, in order to see how this is reflected on the computational power of the systems.

How does the computational power change if we limit the membrane structure to graphs of a restricted form (as a star, a linear tree, a ring, etc.)?

Other possible variants to investigate are the ones considering only one kind of predicate for the edges (i.e., only $\leq$ or only $\geq$).

What happens if the interaction between two conformons is different than addition/subtraction of (part of) the value? (it could be another mathematical operation)

Such study could take advantage from what (i.e., could be done using the techniques) described in [34].

4.4 Modelling with conformons

As indicated in Section 4.2 one of the trends in research in Membrane Computing is to use P systems to model (biological) processes.

It is not surprising that a computationally complete computing device can model a (biochemical) process (except quantum phenomena [28]). Things become more interesting if the simulation of a process brings to the confirmation or falsification of some laws related to that process. So in modelling we should aim at this.

It is also obvious that all computationally complete devices can model the same family of processes. In the moment a device is chosen as tool to model a process, then the motivations for this particular choice have to be explained.

Compared to other variants of P systems, conformon-P system have a rather simple definition that allows a better analysis of the information process. Their simple definition permits anyhow to create sub-systems (called *modules* in [33]) able to perform complex functions that can be regarded as atomic ones.

Suggestion for research: Create different conformon-P systems modelling one particular process and then study how the different conformon-P systems confirm or falsify laws related to the considered process.

Trying to relate the confirmation or falsification of the laws to some properties (in the information process, in the underlying graph, etc.) of the different conformon-P systems.

The just suggested study could be an initial phase for:

Suggestion for research: Compare different processes in terms of the properties that the conformon-P systems simulating them have.

The result of such study could be of the kind: process A and process B are similar because (more than belonging to a certain computability class) they can be modelled by conformon-P systems having these properties (again: in the information process, in the underlying graph, etc.).

It is possible to consider conformon-P systems (and any other variant of P system) placed in a grid as the one depicted in Figure 5. Once a grid has been created, then it is possible to study its evolution as the movement of some conformon, the number of conformon in a (group of) membrane, the application of a rule, etc.

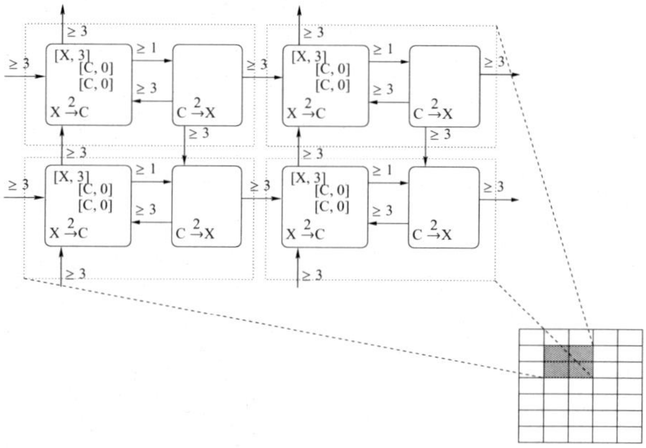

Figure 5: A grid based on conformon-P systems

Once this is done studies similar to the ones present for Cellular Automata [45] can be pursued. For instance:

Suggestion for research: What are the properties of conformon-P systems placed in a grid generating specific patterns?

What has to be changed to a conformon-P systems placed in a grid in order to generate a pattern?

Why such change generate new patterns?

5 Computing by observation

In [16] membrane systems are the basic device underlying a new paradigm of computation: *computing by observation*.

In computing by observation the outcome of a computation is not directly provided by a computing device, instead it is provided by the interaction of three components: a *basic system*, an *observer*, and a *decider*.

Each configuration present in a computation of the basic system is 'analysed' by the observer. As the result of each 'analysis' a symbol is emitted by the observer. This sequence of symbols is the input of the decider that can give a binary output: 'yes' or 'no'. This is sketched in Figure 6.

Computing by observation originates by noticing that "when biologists describe an experiment, most of the times, they are not interested only in the last phase of it. Instead they observe all the process, how it changes, and decide the success/failure of the experiment according to the sequence of changes"[†]

[†]extract from a discussion with M. Cavaliere

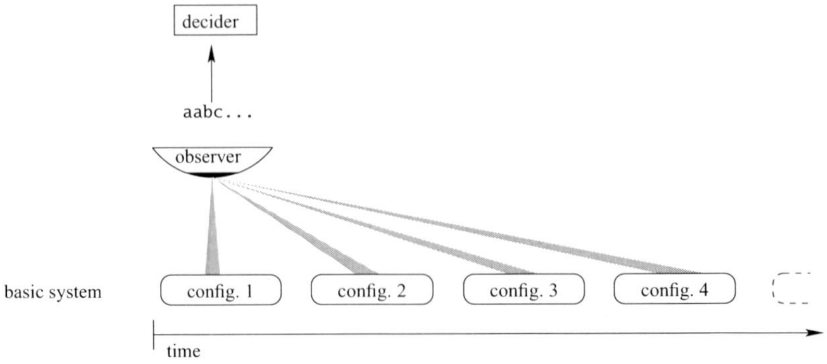

Figure 6: Sketch of computing by observation

In a series of paper [3, 17] it is proved that the superposition of (computationally) simple systems in the just described fashion, leads to a system with an higher computational power.

Computation by observation is definitely an interesting idea that needs to be investigated further.

Suggestion for research: Investigate computing by observation in more details.

What happens when changing the computational power of the three components present in it?

Is it possible to express all the levels of the Chomsky hierarchy?

Is it possible to go beyond Turing power?

6 Final remarks

In this paper we gave a short overview on Molecular Computing, with a particular emphasis on Membrane Computing.

We suggested some direction of research concerning computational aspects of conformon-P systems (placed in a grid), modelling of conformon-P systems (placed in a grid), and computational aspects of computing by observation.

We tried not to use a technical language in the hope to make this paper readable to a wider audience, moreover we list an extensive bibliography for further reading.

References

[1] L. M. Adleman. Molecular computation of solutions to combinatorial problems. *Science*, 266:1021–1024, November 11, 1994.

[2] L. M. Adleman. On constructing a molecular computer. In Lipton and Baum [64], pages 1–22.

[3] A. Alhazov and M. Cavaliere. Computing by observing bio-systems: The case of sticker systems. In Ferretti et al. [27], pages 1–13.

[4] M. Amos. *Theoretical and Experimental DNA Computation.* Natural computing series. Springer-Verlag, Berlin, Heidelberg, New York, 2005.

[5] R. Barish, P.W.K. Rothemund, and E. Winfree. Algorithmic self-assembly of a binary counter using DNA tiles. In Carbone et al. [14].

[6] E. A. Baum. A DNA associative memory potentially larger than the brain. In Lipton and Baum [64], pages 23–28.

[7] D. Beaver. Molecular computing. Technical Report TR95-001, Penn State University, January 31 1995.

[8] D. Beaver. Universality and complexity of molecular computation. Unpublished, 1996.

[9] Y. Benenson, T. Paz-Elizur, R. Adar, E. Keinan, Z. Livneh, and E. Shapiro. Programmable and autonomous computing machine made of biomolecules. *Nature*, 414(1):430–434, November 22, 2001.

[10] C. H. Bennett. Logical reversibility of computation. *IBM Journal of Research and Development*, 17:525–532, 1973.

[11] C. H. Bennett. Thermodynamics of computation – a review. *International Journal of Theoretical Physics*, 21:905–940, 1982.

[12] D. Boneh, C. Dunworth, R. J. Lipton, and J. Sgall. On the computational power of DNA. *Discrete Applied Mathematics*, 71(1-3):79–94, 1996.

[13] R. S. Braich, N. Chelyapov, C. Johnson, P. W. K. Rothemund, and L. Adleman. Solution to a 20-variable 3-SAT problem on a DNA computer. *Science*, 296(5567):499–502, 2002.

[14] A. Carbone, M. Daley, L. Kari, I. McQuillan, and N. Pierce, editors. *Preproceedings 11th International Workshop on DNA Computing, DNA 10, London, Ontario, Canada, June 6-9, 2005*. The university of Wester Ontario, 2005.

[15] M. Cavaliere, N. Jonoska, S. Yogev, R. Piran, E. Keinan, and N. C. Seeman. Biomolecular implementation of computing devices with unbounded memory. In Ferretti et al. [27].

[16] M. Cavaliere and P. Leupold. Evolution and observation - A new way to look at membrane systems. In C. Martín-Vide, G. Mauri, G. Păun, G. Rozenberg, and A. Salomaa, editors, *Workshop on Membrane Computing*, volume 2933 of *Lecture Notes in Computer Science*, pages 70–87. Springer-Verlag, Berlin, Heidelberg, New York, 2003.

[17] M. Cavaliere and P. Leupold. Evolution and observer: A non-standard way to generate formal languages. *Theoretical Computer Science*, 321(2-3):233–248, 2004.

[18] J. Chen, R. Deaton, and Y.-Z. Wang. A DNA-based memory with in vitro learning and associative recall. In *Preliminary Proceedings of DNA Computing, 9th international Workshop on DNA-Based Computers, DNA 2003, Madison, Wisconsin, USA, 1-4 June 2003*, pages 127–136, 2003.

[19] G. Ciobanu, Gh. Păun, and M.J. Perez-Jimenez, editors. *Applications of Membrane Computing*. Springer-Verlag, Berlin, Heidelberg, New York, 2005.

[20] E. Csuhaj-Varjú and G. Vaszil. P automata or purely communicating accepting P systems. In Păun et al. [83], pages 219–233.

[21] R. Deaton, R. C. Murphy, J. A. Rose, M. H. Garzon, D. R. Franceschetti, and S. E. Stevens, Jr. A DNA based implementation of an evolutionary search for good encodings for DNA computation. In *IEEE International Conference on Evolutionary Computation*, pages 267–271, 1997. Indiana University, Indianapolis, Illinois, USA.

[22] A. Ehrenfeucht, T. Hariu, I. Petre, D. M. Prescott, and G. Rozenberg. Formal systems for gene assembly in ciliates. *Theoretical Computer Science*, 292:199–219, 2003.

[23] A. Ehrenfeucht, T. Harju, I. Petre, D. M. Prescott, and G. Rozenberg. *Computation in Living Cells. Gene Assembly in Ciliates*. Springer-Verlag, Berlin, Heidelberg, New York, 2003.

[24] A. Ehrenfeucht, T. Harju, I. Petre, and G. Rozenberg. Characterizing the micronuclear gene patterns in ciliates. *Theory of Computing Systems*, 35:501–519, 2002.

[25] A. Ehrenfeucht, H. J. Hoogeboom, G. Rozenberg, and N. van Vugt. Forbidding and enforcing. In Winfree and Gifford [102], pages 195–206.

[26] A. D. Ellington, M. P. Robertson, and J. Bull. In vitro evolution - ribozymes in wonderland. *Science*, 276:546–547, April 1997.

[27] C. Ferretti, G. Mauri, and C. Zandron, editors. *10th International Workshop on DNA Computing, DNA 10, Milan, Italy, June 7-10, 2004, Revised Selected Papers*, volume 3384 of *Lecture Notes in Computer Science*. Springer-Verlag, Berlin, Heidelberg, New York, 2005.

[28] R. P. Feynman. Simulating physics with computers. *International Journal of Theoretical Physics*, 21:467–488, 1982.

[29] R. Freund, G. Lojka, M. Oswald, and G. Păun, editors. *Preproceedings of the 6th International Workshop on Membrane Computing WMC6*. TU Wien, Austria, 2005.

[30] R. Freund and M. Oswald. A short note on analysing P systems with antiport rules. *Bulletin of the European Association for Theoretical Computer Science*, 78:231–236, October 2002.

[31] R. Freund, G. Păun, G. Rozenberg, and A. Salomaa. Watson-Crick finite automata. In H. Rubin and D. H. Wood, editors, *DNA Based Computers III*, volume 48 of *DIMACS: Series in Discrete Mathematics and Theoretical Computer Science*, pages 305–317. American Mathematical Society, 1997.

[32] R. Freund, G. Păun, G. Rozenberg, and A. Salomaa. Bidirectional sticker systems. In R. Altman, A. Dunker, L. Hunter, and T. Klein, editors, *Proceedings of Pacific Symposium on Biocomputing*, pages 535–546. World Scientific Publishing Company, 1998. Kapalua, Maui, January 1998, Hawaii, USA.

[33] P. Frisco. The conformon-P system: A molecular and cell biology-inspired computability model. *Theoretical Computer Science*, 312(2-3):295–319, 2004.

[34] P. Frisco. P systems, Petri nets and Program machines. In Freund et al. [29], pages 336–354.

[35] P. Frisco and R. T. Gibson. A simulator and an evolution program for conformon-P systems. In *TAPS, Workshop on Theory and Applications of P Systems, Timisoara (Romania), September 26-27, 2005*, IEEE Computer Press, 2005.

[36] P. Frisco, C. Henkel, and Sz. Tengey. An algorithm for SAT without an extraction phase. In Carbone et al. [14].

[37] P. Frisco and H. J. Hoogeboom. P systems with symport/antiport simulating counter automata. *Acta Informatica*, 41(2-3):145–170, 2004.

[38] P. Frisco and S. Ji. Conformons-P systems. In M. Hagiya and A. Ohuchi, editors, *Proceedings of DNA Computing, 8th international Workshop on DNA-Based Computers, DNA 2002, Sapporo, Japan, June 10-13, 2002. Revised Papers*, volume 2568 of *Lecture Notes in Computer Science*, pages 291–301. Hokkaido University, Springer Verlag, Berlin, Heidelberg, New York, 2002.

[39] P. Frisco and S. Ji. Towards a hierarchy of info-energy P systems. In Păun et al. [83], pages 302–318.

[40] D. E. Green and S. Ji. The electromechanical model of mitochondrial structure and function. *Molecular Basis of Electron Transport*, pages 1–44, 1972. J. Schultz, B. F. Cameron (eds).

[41] M. A. Gutírrez-Naranjo, M. J. Pérez-Jiménez, A. Riscos-Núñez, and F. J. Romero-Campero. On the power of dissolution in p systems with active membranes. In Freund et al. [29], pages 373–394.

[42] T. Head. Circular suggestions for DNA computing. In A. Carbone, M. Gromov, and P. Prusinkiewicz, editors, *Pattern formation in biology, vision and dynamics*, pages 325–335. World Scientific Publishing Company, 2000.

[43] T. Head, G. Rozenberg, R. S. Bladergroen, C. K. D. Breek, P. H. M. Lommerse, and H. P. Spaink. Computing with DNA by operating on plasmids. *BioSystems*, 57(2):87–93, 2000.

[44] T. Head, M. Yamamura, and S. Gal. Aqueous computing: writing on molecules. In *Proceedings of the Congress on Evolutionary Computation*, pages 1006–1010, 1999.

[45] A. Ilachinski. *Cellular automata. A discrete universe*. World Scientific, 2001.

[46] S. Ji. The Bhopalator: a molecular model of the living cell based on the concepts of conformons and dissipative structures. *Journal of Theoretical Biology*, 116:395–426, 1985.

[47] S. Ji. Free energy and information contents of conformons in proteins and DNA. *BioSystems*, 54:107–214, 2000.

[48] S. Ji. The Bhopalator: an information/energy dual model of the living cell (II). *Fundamenta Informaticae*, 49(1-3):147–165, 2002.

[49] N. Jonoska, S. A. Karl, and M. Saito. Three dimensional DNA structures in computing. In L. Kari, H. Rubin, and D. H. Wood, editors, *DNA based computer 4*, volume 52 of *DIMACS: series in discrete mathematics and theoretical computer science*, pages 143–153, 1998.

[50] N. Jonoska and N. C. Seeman, editors. *DNA Computing: 7th International Workshop on DNA-Based Computers, DNA7, Tampa, FL, USA, June 10-13, 2001. Revised Papers*, volume 2340. Springer Verlag, Berlin, Heidelberg, New York, 2001.

[51] L. Kari. *On insertion and deletion in formal languages*. PhD thesis, University of Turku, Finland, 1991.

[52] L. Kari, J. Kari, and L. Landweber. *Jewels are forever*, chapter Reversible molecular computation in ciliates, pages 353–363. Springer Verlag, Berlin, Heidelberg, New York, 1999. J. Karhumäki and H. Maurer and G. Păun and G. Rozenberg (eds).

[53] L. Kari., G. Păun, G. Rozenberg, A. Salomaa, and S. Yu. DNA computing, sticker systems and universality. *Acta Informatica*, 35:401–420, 1998.

[54] L. Kari and G. Thierrin. Contextual insertions/deletions and computability. *Information and Computation*, 131(1):47–61, November 25 1996.

[55] S. Kashiwamura, M. Yamamoto, A. Kameda, T. Shiba, and A. Ohuchi. Hierarchical DNA memory based on nested PCR. Poster paper at the 8th International Workshop on DNA-Based Computers, DNA 2002, Sapporo, Japan, 10-13 June 2002.

[56] G. Kemeny and I. M. Goklany. Polarons and conformons. *Journal of Theoretical Biology*, 40:107–123, 1973.

[57] G. Kemeny and I. M. Goklany. Quantum mechanical model for conformons. *Journal of Theoretical Biology*, 48:23–38, 1974.

[58] S. N. Krishna. *Languages of P systems. Computability and complexity.* PhD thesis, IIT Madras, India, 2001.

[59] S. N. Krishna and R. Rama. A variant of P systems with active membranes: solving NP-complete problems. *Romaninan Journal of Information Science and Technology*, 2(4):357–367, 1999.

[60] L. Landweber and E. Baum, editors. *DNA Based Computers II*, volume 44 of *DIMACS: Series in Discrete Mathematics and Theoretical Computer Science*. American Mathematical Society, 1996.

[61] L. F. Landweber and L. Kari. *Evolution as computation*, chapter Universal molecular computation in ciliates, pages 257–274. In Landweber and Winfree [62], 1999.

[62] L. F. Landweber and E. Winfree, editors. *Evolution as computation*. Natural computing series. Springer Verlag, Berlin, Heidelberg, New York, 1999.

[63] R. J. Lipton. Using DNA to solve NP-complete problems. *Science*, 268:542–545, April 28, 1995.

[64] R. J. Lipton and E. B. Baum, editors. *DNA based computers*, volume 27 of *DIMACS: Series in Discrete Mathematics and Theoretical Computer Science*. American Mathematical Society, 1995.

[65] V. Manca. DNA and membrane algorithms for SAT. *Fundamenta Informaticae*, 49(1-3):205–221, 2002.

[66] V. Manca and C. Zandron. A DNA algorithm for 3-SAT(11, 20). In Jonoska and Seeman [50], pages 172–181.

[67] C. Martín-Vide, G. Păun, and G. Rozenberg. Membrane systems with carriers. *Theoretical Computer Science*, 270:779–796, 2002.

[68] C. Martín-Vide, G. Păun, G. Rozenberg, and A. Salomaa. *Computing with Bio-Molecules. Theory and Experiments*, chapter Universality results for finite H systems and for Watson-Crick finite automata, pages 200–220. Springer Verlag, Berlin, Heidelberg, New York, 1998.

[69] J. S. McCaskill. Optically programming DNA computing in microflow reactors. *BioSystems*, 59(2):125–138, February 2001.

[70] J. S. McCaskill and P. Wagler. From reconfigurability to evolution in construction systems: spanning the electronic, microfuidic and biomolecular domains. In R. W. Hartenstein and H. Grunbacher, editors, *Field-Programmable Logic and Applications. The Roadmap to Reconfigurable Computing: 10th International Conference, FPL 2000, Villach, Austria, August 27-30, 2000. Proceedings*, volume 1896 of *Lecture Notes in Computer Science*, pages 286–299. Springer Verlag, Berlin, Heidelberg, New York, 2000.

[71] Z. Michalewicz. *Genetic algorithms + data structures = evolution programs*. Springer-Verlag, Berlin, Heidelberg, New York, 1999.

[72] M. L. Minsky. *Computation: Finite and Infinite Machines*. Automatic computation. Prentice-Hall, 1967.

[73] J. J. Mulawka, P. Wasiewicz, and K. Pietak. Virus-enhanced genetic algorithms inspired by DNA computing. *Lecture Notes in Computer Science*, 1609:527–537, 1999.

[74] H. Nakagawa, K. Sakamoto, and Y. Sakakibara. Development of an *in vivo* computer based on *escherichia coli*. In Carbone et al. [14].

[75] A. Obtulowicz. Deterministic P systems for solving SAT. *Romaninan Journal of Information Science and Technology*, 4(1-2):195–202, 2001.

[76] Q. Ouyang, P. D. Kaplan, S. Liu, and A. Libchaber. DNA solution of the maximal clique problem. *Science*, 278:446–449, 1997.

[77] C. H. Papadimitriou. *Computational complexity*. Addison-Wesley Pub. Co., 1994.

[78] A. Păun and G. Păun. The power of communication: P systems with symport/antiport. *New Generation Computing*, 20(3):295–306, 2002.

[79] G. Păun. Computing with membranes. *Journal of Computer and System Sciences*, 1(61):108–143, 2000.

[80] G. Păun. P systems with active membranes: attacking NP complete problems. *Journal of Automata, Languages and Combinatorics*, 1(6):75–90, 2001.

[81] G. Păun. *Membrane Computing. An Introduction*. Springer-Verlag, Berlin, Heidelberg, New York, 2002.

[82] G. Păun and G. Rozenberg. Sticker systems. *Theoretical Computer Science*, 205, 1998.

[83] G. Păun, G. Rozenberg, A. Salomaa, and C. Zandron, editors. *Membrane Computing: International Workshop, WMC-CdeA 2002, Curtea de Arges, Romania, August 19-23, 2002. Revised Papers*, volume 2597 of *Lecture Notes in Computer Science*. Springer-Verlag, Berlin, Heidelberg, New York, 2002.

[84] G. Păun, Y. Sakakibara, and T. Yokomori. P systems on graphs of restricted forms. *Publicationes Mathematicae Debrecen*, 2001.

[85] G. Păun and T. Yokomori. Membrane computing based on splicing. In Winfree and Gifford [102], pages 217–232.

[86] J. H. Reif. Parallel molecular computation: models and simulations. In *Proceedings of the 7th Annual ACM Symposium on Parallel Algorithms and Architectures (SPAA95), Santa Barbara, June 1995*, pages 213–223, 1995.

[87] J. H. Reif, T. H. LaBean, M. Pirrug, V. S. Rana, B. Guo, C. Kingsford, and G. S. Wickham. Experimental construction of a very large scale DNA database with associative search capability. In Jonoska and Seeman [50], pages 231–247.

[88] P. W. K. Rothemund. A DNA and restriction enzyme implementation of Turing machines. In Lipton and Baum [64], pages 75–120.

[89] S. Roweis, E. Winfree, R. Burgoyne, N. V. Chelyapov, M. F. Goodman, P. W. K. Rothemund, and L. M. Adleman. A sticker based model for DNA computation. In Landweber and Baum [60].

[90] K. Sakamoto, H. Gounzu, K. Komiya, D. Kiga, S. Yokoyama, T. Yokomori, and M. Hagiya. Molecular computation by DNA hairpin formation. *Science*, 288:1223–1226, May 19, 2000.

[91] R. Schulman and E. Winfree. Programmable control of nucleation for algorithmic self-assembly. In Ferretti et al. [27].

[92] D. Soloveichik and E. Winfree. Complexity of self-assembled shapes. In Ferretti et al. [27].

[93] M. Soreni, S. Yogev, E. Kossoy, Y. Shoham, and Eh. Keinan. Parallel biomolecular computation on surfaces with advanced finite automata. *Journal American Chemical Society*, 127(11):3935–3943, 2005.

[94] D. van Noort, Frank-Ulrich Gast, and J. S. McCaskill. DNA computing in microreactors. In Jonoska and Seeman [50], pages 33–45.

[95] N. van Vugt. *Models of molecular computing*. PhD thesis, LIACS, Universiteit Leiden, The Netherlands, 2002.

[96] M. V. Volkenstein. The conformon. *Journal of Theoretical Biology*, 34:193–195, 1972.

[97] R. Weiss, S. Basu, S. Hooshangi, A. Kalmbach, D. Karig, R. Mehreja, and I. Netravali. Genetic circuit building blocks for cellular computation, communications, and signal processing. *Natural Computing*, 2(1):43–84, 2003.

[98] K. Wickiser, S. Chalamish-Chohen, S. Narashimhan, B. Sawichi, D. C. Ward, and R. R. Breaker. Oligonucleotide sensitive hammerhead ribosomes as logic gates. In *Preproceedings DNA Computing: 8th International Workshop on DNA-Based Computers, DNA8 Sapporo, Japan, June 10-13*, 2002. page 19.

[99] E. Winfree. On the computational power of DNA annealing and ligation. In Lipton and Baum [64], pages 199–210.

[100] E. Winfree. *Algorithmic Self-Assembly of DNA*. PhD thesis, California Institute of Technology, Pasadena, California, May 19 1998.

[101] E. Winfree, T. Eng, and G. Rozenberg. String tile models for DNA computing by self-assembly. In A. Condon and G. Rozenberg, editors, *DNA Computing*, volume 2054 of *Lecture Notes in Computer Science*, pages 63–88. Springer Verlag, Berlin,

Heidelberg, New York, 2000. 6th International Workshop on DNA-Based Computers, DNA 2000, Leiden, The Netherlands, June 13-17, 2000, Revised Papers.

[102] E. Winfree and D. K. Gifford, editors. *DNA Based Computers V*, volume 54 of *DIMACS: Series in Discrete Mathematics and Theoretical Computer Science*. American Mathematical Society, 1999.

[103] E. Winfree, F. Lin, L. A. Wenzler, and N. C. Seeman. Design and self-assembly of two-dimensional DNA crystals. *Nature*, 394(6693):539–545, August 6 1998.

[104] E. Winfree, X. Yang, and N. C. Seeman. Universal computation via self-assembly of DNA: some theory and experiments. In Landweber and Baum [60].

[105] D. H. Wood, H. Bi, S. O. Kimbrough, D. Wu, and J. Chen. DNA starts to learn poker. In Jonoska and Seeman [50], pages 92—103.

[106] C. Zandron. P-systems web page: `http://psystems.disco.unimib.it`.

Pierluigi Frisco
Dept. of Comp. Sci., School of Eng., C.S., and Math.
University of Exeter, Harrison Building
North Park Road, EX4 4QF Exeter, U.K.
email: `p.frisco@ex.ac.uk`

Formal Language Aspects of Natural Computing– 2006, pp. 103–112

An Overview of Recent Results in Contextual Grammars for MCS Formalisms

K. Lakshmanan

Abstract

Contextual grammars were introduced by Marcus in 1969 [15] as 'intrinsic grammars', without the auxiliary symbols, based on the fundamental linguistic operation of inserting words according to the given selection procedure. Relevance of contextual grammars with natural computing can be found in [4, 6, 7, 8, 9]. In this paper, rather than focusing our attention to contextual grammars in relevance to natural computing, we focus our aim to the applications of contextual grammars to the formalisms of natural languages. We report some new classes of contextual grammars which are recently introduced in order to show their relevance to natural languages in view of the feature of mildly context sensitiveness, since mildly context sensitive (MCS) formalisms are considered to be an appropriate syntactical description for natural languages.

1 Introduction

Contextual grammars produce languages starting from a finite set of *axioms* and adjoining *contexts*, iteratively, according to the *selector* present in the current sentential form. As introduced in [15], adjoining the contexts are done at the ends of the strings. Internal contextual grammars were introduced by Păun and Nguyen in 1980 [20], where the contexts are adjoined to the selector strings appeared as substrings of the derived string. More precisely, given a word x (which is a subword of the derived word w) and a context (u, v), we adjoin the strings u to the left of x and v to the right of x in w and the language is generated iteratively in this fashion. Later on, many variants of contextual grammars were introduced consequently by imposing restriction to the contexts viz., *leftmost, one-sided*, and in choosing the selectors viz., *maximal use of selector*, *depth-first*. The operation of adjoining and erasing of contexts are counterpart of *insertion* and *deletion* operation where given a pair of words (u, v) (which are subwords of the derived word w), the insertion or deletion of a string x is performed between u and v in w.

Now, we will see some results on contextual grammars and insertion deletion grammars in relevance to natural computing. In [7], a new model of P systems called *hybrid P systems* was considered by combining the context-free and contextual rules. In [8], P systems with string objects having contextual rules as the control structure was considered and the model was referred as *contextual P systems*. P systems with similar feature were also investigated in [5]. In [9], tissue P systems with context-free and contextual rules were considered. The computational completeness and the other generative power of these systems were investigated in the above mentioned papers. Also, some NP-complete problems (Hamiltonian path problem and satisfiability problem) were solved in polynomial time but with exponential space complexity in the framework of contextual P systems.

Insertions and deletions of small circular DNA strands into long linear DNA strands are phenomena that happen frequently in nature and thus constitute an attractive paradigm for biomolecular computing. In [4], contextual insertion and deletion for biomolecular was considered and in [2], a new model for DNA-based computation that involves circular as well as linear molecules with insertion and deletion operation was considered. P systems with insertion-deletion rules were considered in [6] and tissue P systems with insertion-deletion rules were considered in [13]. The generative power of these systems were investigated in comparison with $CF, MAT, ET0L, E0L, RE$. There are also other investigations made in recent times, for instance, in [14], the above said NP-complete problems were solved using tissue P systems with insertion-deletion rules.

But, in this paper we do not present any results on contextual grammars regarding natural computing, rather we heed to another important application of contextual grammars, which is natural languages. A typical class of languages is said to have MCS formalism if the languages satisfy certain properties of *mildly context sensitiveness*, which are:

1. The languages should contain the following three basic non-context-free constructions: (i) *multiple agreements:* $L = \{a^n b^n c^n | n \geq 1\}$, (ii) *crossed dependencies:* $L = \{a^n b^m c^n d^m | n, m \geq 1\}$, and (iii) *marked duplication:* $L = \{wcw | w \in \{a, b\}^*\}$.

2. All the languages are *parsable in polynomial time*. In other words, the *membership problem* can be solved in polynomial time algorithm.

3. The languages contain *semilinear* languages only.

As mildly context sensitive (MCS) formalisms are considered to be an appropriate model for the description of natural languages, it is important in formal language theory as well as in natural language processing to obtain certain classes of languages which satisfy the above said MCS property in order to have the MCS formalism.

Even though contextual grammars were introduced to give an appropriate model for natural languages, the basic variants of internal contextual languages failed to contain the non-context-free constructions itself. So, new classes of grammars were introduced to tackle this problem, for instance, maximal use of selector [16]. Though they generate the basic non-context-free languages, they contain non-semilinear languages and also their membership problem either remains open or is solvable with exponential time complexity [1].

But, during this last decade, some attempts have been made to introduce some variants of contextual grammars by restricting the *selector* and *contexts* to obtain certain specific classes of contextual languages which satisfy the *MCS* property [3, 11, 12]. In [3], Ilie considered a new variant called *local* contextual grammars, in which the semilinear property was left unsolved. In [12], kudlek et al. considered a family of *p-dimensional external* contextual grammars, and showed that these languages satisfy the properties of mildly context-sensitive languages. But in these p-dimensional contextual grammars the derivation takes place in *external* mode. In [11], a variant motivated from local contextual grammar called *absorbing right context grammars* was considered in order to resolve the semilinear problem. This class of languages was shown to satisfy all the properties of MCS formalisms. In this paper, we will see the following variants of contextual grammars and analyze their languages with the above said MCS properties: *internal, depth-first, lengthwise depth-first, maximal, maximal lengthwise depth-first, local, absorbing right context.*

2 Classes of Contextual Grammars

We assume that the readers are familiar with the basic notions of formal language theory. For the basic details on formal language theory, we refer to [21].

Now, we shall proceed to define the classes of contextual grammars. There are three basic classes of contextual grammars which are *external, internal* and *total*. As the results are based on internal contextual grammar only, we introduce this grammar only in this paper and hereafter we simply refer it as grammar. An *internal contextual grammar* is a construct $G = (V, A, (S_1, C_1), ..., (S_m, C_m))$, $m \geq 1$, where V is an alphabet, $A \subseteq V^*$ is a finite set called the set of *axioms*, $S_i \subseteq V^*$, $1 \leq i \leq m$, are the sets of *selectors*, $C_i \subseteq V^* \times V^*$, C_i finite, $1 \leq i \leq m$, are the sets of *contexts*.

The *modular presentation* [19] of a contextual grammar is given as $G = (V, A, P)$ where V, A are as defined above and P is the finite set of selector-context rules of the form $(S_1, C_1), \ldots, (S_m, C_m)$. The usual derivation in the *internal mode* is defined as $x \Longrightarrow_{in} y$ iff

$$x = x_1x_2x_3,\ y = x_1ux_2vx_3,\ \text{for } x_1, x_2, x_3 \in V^*,\ x_2 \in S_i,\ (u, v) \in C_i,\ 1 \leq i \leq m.$$

The language generated by the above grammar G is given as

$$L_{in}(G) = \{x \in V^* \mid w \Longrightarrow_{in}^{*} x,\ w \in A\},$$

where $\Longrightarrow_{in}^{*}$ is the reflexive, transitive closure of the relation $\Longrightarrow_{in}$. If all the selectors $S_1, S_2, \ldots, S_n$ are in a family F of languages, then we say that the grammar G is with F selection or selector.

Example 1. *Consider the grammar* $G_1 = (\{a, b\}, ab, \{(ab, (a, b))\}$.
Any derivation in G_1 *of internal mode is*

$$^{\downarrow}ab^{\downarrow} \Longrightarrow_{in} \underline{a}^{\downarrow}ab^{\downarrow}\underline{b} \Longrightarrow_{in} aa\underline{a}^{\downarrow}ab^{\downarrow}\underline{b}bb \Longrightarrow_{in} aaa\underline{a}^{\downarrow}ab^{\downarrow}\underline{b}bbb \Longrightarrow_{in}^{*} a^nb^n.$$

In the derivation, the string between the two down arrows indicate the selector for the next derivation step and the underlined symbols denotes the contexts introduced in the current derivation step. It is easy to see that, $L_{in}(G_1) = \{a^nb^n \mid n \geq 1\}$. *Note the the selector ab is of finite type.*

Though, internal contextual grammars generate the 'crossed dependencies' languages, they cannot generate 'multiple agreements' and 'marked duplication'. Also, the membership algorithm takes exponential time complexity for this family of languages [1]. Besides, the languages generated by these grammars with finite selectors contain non semilinear languages. So, the family of languages generated by internal contextual grammars does not satisfy the properties needed for the MCS formalisms.

Next, we shall introduce depth-first contextual grammars. *Depth-first* contextual grammars were introduced in [18]. In depth-first contextual grammars, at each derivation step (except the first one), at least one of the words of the context u or v introduced in the previous derivation step must be a subword of the currently used selector.

Example 2. *Consider the following grammar*

$$G_2 = (\{a, b, c, d\}, abcd, (a^+b^+c^+, (a, c)), (b^+c^+d^+, (b, d))).$$

Any derivation in G_2 *of depth-first mode is*

$$^{\downarrow}abc^{\downarrow}d \Longrightarrow_{df}\ ^{\downarrow}\underline{a}abc\underline{c}^{\downarrow}d \Longrightarrow_{df} \underline{a}aa^{\downarrow}bc\underline{c}cd^{\downarrow} \Longrightarrow_{df} aaa\underline{b}bcccd\underline{d} \Longrightarrow_{df}^{*} a^nb^mc^nd^m.$$

Note that, at each derivation step (except the first one) the selector contains one of the contexts which was introduced in the previous derivation step. It is easy to see that, $L_{df}(G_2) = \{a^nb^mc^nd^m \mid n, m \geq 1\}$. *Note that here the selectors are regular type.*

The restricted version of depth-first contextual grammars called *lengthwise depth-first* contextual grammars were introduced in [10], where at each derivation step (except the first one), the selector must contain the context u or v (introduced in the previous derivation step) whichever is of maximal length. When the lengths of u and v are equal length, the selector is chosen non-deterministically (which contains u or v).

Though, the above two families of grammars generate crossed dependencies and marked duplication, they are not able to generate multiple agreements language. Also, the membership and semilinear problem are still left open.

Now, we shall introduce the grammars which can generate the basic non-context-free languages. Given a contextual grammar G, we define the *maximal use of selectors* in the following way.

$$x \Longrightarrow_m y \text{ iff } x = x_1x_2x_3,\ y = x_1ux_2vx_3, \text{ for } x_2 \in S_i,\ (u_i, v_i) \in C_i,\ 1 \leq i \leq m,$$
$$\text{and there are } \textbf{no } x'_1, x'_2, x'_3 \in V^*, \text{ such that } x = x'_1x'_2x'_3,\ x'_2 \in S_i, \text{ and}$$
$$x'_2 \text{ contains } x_2 \text{ (i.e. } |x'_2| > |x_2|).$$

This means that, no strict superword of the *selector* can be used by the same production. We say that $\Longrightarrow_{max}$ is a derivation in the *maximal* mode. We now introduce the next variant, maximal depth-first grammars which is defined by combining the maximal and the depth-first variants. Formally, given a contextual grammar G, a *maximal depth-first derivation* (denoted by *mdf*) in G is a derivation $w_1 \Longrightarrow_{mdf} w_2 \Longrightarrow_{mdf} \cdots \Longrightarrow_{mdf} w_n,\ n \geq 1$, where

(i) $w_1 \in A$, $w_1 \Longrightarrow_m w_2$ in the maximal way,

(ii) For each $i = 2, 3, \ldots, n-1$, if $w_{i-1} = z_1z_2z_3$, $w_i = z_1uz_2vz_3$ ((u, v) is the context adjoined to w_{i-1} in order to get w_i), then $w_i = x_1x_2x_3$, $w_{i+1} = x_1sx_2tx_3$, such that $x_2 \in S_j$, $(s, t) \in C_j$, for some j, $1 \leq j \leq m$, and x_2 contains one of the contexts u **or** v as a substring. That is, at every derivation step, the selector $x_2 \in S_j$ contains one of the contexts u **or** v which was adjoined in the previous derivation step (satisfying depth-first condition).

(iii) For each $i = 2, 3, \ldots, n-1$, if $w_i \Longrightarrow_{df} w_{i+1}$ then there will be **no** other derivation in G with $w_i \Longrightarrow_{df} w'_{i+1}$ such that $w_i = x'_1x'_2x'_3$, $x'_2 \in S_j$ and $|x'_2| > |x_2|$ where $x_2 \in S_j$. That is, at every derivation step, w_i should have **no** other subword x'_2 with $x'_2 \in S_j$ of greater length than $x_2 \in S_j$. (satisfying maximal mode condition).

In a similar way, we can define *maximal lengthwise depth-first contextual grammars* (denoted by *mldf*), where, at each derivation step, the selector contains one of the contexts u or v whichever is of maximum length and the selector is *maximal*.

The next variant we define is local contextual grammars. Given a contextual grammar G, we define the *local* mode in the following way. for $z \in A$, $z \Longrightarrow_{in} x$ such that $z = z_1z_2z_3$, $x = z_1\underline{u}z_2\underline{v}z_3$, $z_2 \in S_i$, $(u, v) \in C_i$, for $z_1, z_2, z_3 \in V^*$, $1 \leq i \leq n$, then $x \Longrightarrow_{loc} y$ is called local w.r.t. $z \Longrightarrow x$, iff we have $u = u'u''$, $v = v'v''$, $u'u'', v', v'' \in V^*$, for $x = z_1u'^{\downarrow}u''z_2v'^{\downarrow}v''z_3$, $y = z_1u'\underline{s}u''z_2v'\underline{t}v''z_3$ for $u''z_2v' \in S_j$, $(s, t) \in C_j$ for $1 \leq j \leq n$.

The derivation is called local because at each derivation step, the contexts for the further derivation are adjoined inside or near the contexts which are introduced in the earlier derivation step. When the maximal restriction is included with this local variant, the grammar is said to be *maximal local* (denoted by *mloc*). That is, the selector for the next derivation satisfies the local condition as well as maximal condition. Note that in such case, the first derivation $z \Longrightarrow x$ should be performed in maximal mode.

Now, we introduce a variant which was obtained by imposing few more restrictions to the above local contextual grammar and is called *absorbing right contextual grammars*. In this variant, the selector (say y_{i+1}) for the next derivation (step) is obtained by adjoining the first half v_i' of the current right context ($v_i = v_i'v_i''$) to the current selector y_i. That is, $y_{i+1} = y_i v_i'$, $y_i \in S_j, y_i, v_i' \in V^*$. More formally, given contextual grammar G, the mode of derivation step is as follows: if we have the derivation step $z \Longrightarrow x$ such that $z = z_1z_2z_3$, z_1, z_2, $z_3 \in V^*$, $z_2 \in S_i$, $(u, v) \in C_i$, for some $1 \leq i \leq m$, and $x = z_1uz_2vz_3$, then a derivation $x \Longrightarrow y$ is called *absorbing right context* with respect to the derivation $z \Longrightarrow x$, denoted by $z \Longrightarrow x \Longrightarrow_{arc} y$, iff the following holds:

1. Assume that $v = v'v''$, where $v', v'' \in V^*$, and $|v'| = \lceil \frac{|v|}{2} \rceil$, $|v''| = \lfloor \frac{|v|}{2} \rfloor$. If $v = a_1a_2 \ldots a_r$, then $v' = a_1 \ldots a_{\lceil \frac{r}{2} \rceil}$, $v'' = a_{\lfloor \frac{r}{2} \rfloor} \ldots a_r$. (That is the right adjoined context is decomposed into two halves.)

2. $x = x_1x_2x_3$ for $x_1 = z_1u$, $x_2 = z_2v'$, $x_3 = v''z_3$, $x_2 \in S_j$, $(s, t) \in C_j$, for some $1 \leq j \leq n$ and $y = x_1sx_2tx_3$. (That is the right half context v' is absorbed with the previous selector x_2.)

From the above derivation, it is clear that the selector absorbs the right context (in part and sometimes in full, when $|v_i| = 1$) introduced in the earlier derivation, and hence one calls *absorbing right context* grammars, denoted by *arc*. The class of languages $L_\alpha(G)$, for G working in the *max*, *mloc*, *arc*, *mdf*, *mldf* mode with F selection is denoted by $ICC_{max}(F)ICC_{ml}(F)$, $ICC_{arc}(F)$, $ICC_{mdf}(F)$, $ICC_{mldf}(F)$, respectively.

3 Analyzing Mildly Context Sensitiveness

In this section, we will see some results of the above introduced variants from the point of mildly context sensitiveness. We will how much these variants are feasible to the MCS formalisms.

3.1 Generation of Non-Context-Free Languages

We show in this section that all the three basic non-context-free constructions in natural languages, that is *multiple agreements*, *crossed dependencies* and *marked duplication*, can be realized using *max*, *mloc*, *arc*, *mdf*, *mldf* with regular selection.

Lemma 1. *(multiple agreements)*
$L_1 = \{a^nb^nc^n | n \geq 1\} \in ICC_\alpha(REG)$, $\alpha \in \{max, mloc, arc, mdf, mldf\}$.

Proof: First, we prove the result for *max* and *mloc* grammars. Consider the following grammar

$$G_1 = (\{a, b, c\}, \ abc, \ \{(b^+, (ab, c))\}).$$

Any derivation in $\beta \in \{max, mloc\}$ mode of G_1 is as follows:

$$a^{\downarrow}bc^{\downarrow} \Longrightarrow_\beta a\underline{a^{\downarrow}b}b\underline{c}c^{\downarrow} \Longrightarrow_\beta aa\underline{a^{\downarrow}b}bb\underline{c}cc^{\downarrow} \Longrightarrow_\beta aaa\underline{a^{\downarrow}b}bbb\underline{c}ccc^{\downarrow} \Longrightarrow_\beta^* a^nb^nc^n.$$

Note that for the case of maximal local mode, $u' = a, u'' = b, v' = \alpha, v'' = c$.

For arc grammars, consider the grammar $G'_1 = (\{a, b, c\}, abc, \{(b^+, (a, bc))\})$.
Any derivation in *arc* mode of G'_1 is as follows:

$$a^{\downarrow}b^{\downarrow}c \Longrightarrow_{arc} a\underline{a}^{\downarrow}\underline{bb}^{\downarrow}\underline{cc} \Longrightarrow_{arc} aa\underline{a}^{\downarrow}bb\underline{b}^{\downarrow}\underline{ccc} \Longrightarrow_{arc} aaa\underline{a}^{\downarrow}bbb\underline{b}^{\downarrow}cccc \Longrightarrow^{*}_{arc} a^n b^n c^n.$$

For mdf and $mldf$ grammars, consider the following grammar,

$$G''_1 = (\{a, b, c\}, abc, \{(a^+b^+, (a, bc)), (b^+c^+, (ab, c))\}).$$

Any derivation in *mdf* mode of G''_1 is as follows:

$$^{\downarrow}ab^{\downarrow}c \Longrightarrow_{mdf} {}^{\downarrow}\underline{aabb}^{\downarrow}\underline{cc} \Longrightarrow_{mdf} \underline{aa}a^{\downarrow}bb\underline{bccc}^{\downarrow} \Longrightarrow_{mdf} aaa\underline{abbbbcccc} \Longrightarrow^{*}_{mdf} a^n b^n c^n.$$

Any derivation in $mldf$ mode of G''_1 is

$$a^{\downarrow}bc^{\downarrow} \Longrightarrow_{mldf} {}^{\downarrow}\underline{aabb}^{\downarrow}c\underline{c} \Longrightarrow_{mldf} \underline{aa}a^{\downarrow}bb\underline{bccc}^{\downarrow} \Longrightarrow_{mldf} aaa\underline{abbbbcccc} \Longrightarrow^{*}_{mldf} a^n b^n c^n.$$

□

Lemma 2. *(crossed agreements)*
$L_2 = \{a^n b^m c^n d^m | n, m \geq 1\} \in ICC_{\alpha}(REG),\ \alpha \in \{max, mloc, arc, mdf, mldf\}$.

Proof: Consider the grammar $G_2 = (\{a, b, c, d\}, abcd, \{(b^+, (a, c)), (c^+, (b, d))\})$. Any derivation in max mode of G_2 is given by

$$a^{\downarrow}b^{\downarrow}cd \Longrightarrow_{max} a\underline{a}^{\downarrow}b^{\downarrow}\underline{c}cd \Longrightarrow_{max} aa\underline{a}b^{\downarrow}\underline{c}cc^{\downarrow}d \Longrightarrow_{max} aaab\underline{b}ccc\underline{d}d \Longrightarrow^{*}_{max} a^n b^m c^n d^m.$$

Consider the grammar $G'_2 = (\{a, b, c, d\}, abcd, (b^+c^+, \{(a, c), (b, d)\}))$.
Any derivation in β, $\beta \in \{mloc, mdf, mldf\}$ mode of G'_2 is as follows:

$$a^{\downarrow}bc^{\downarrow}d \Longrightarrow_{\beta} a\underline{a}^{\downarrow}bc\underline{c}^{\downarrow}d \Longrightarrow_{\beta} aa^{\downarrow}\underline{b}bcc^{\downarrow}\underline{d}d \Longrightarrow_{\beta} aa^{\downarrow}\underline{b}bbcc^{\downarrow}\underline{d}dd \Longrightarrow_{\beta}$$
$$a\underline{a}a^{\downarrow}bbbcc\underline{c}^{\downarrow}ddd \Longrightarrow_{\beta} \cdots \Longrightarrow_{\beta} a^n b^m c^n d^m, \text{ for some } n, m \geq 1.$$

It is easy to see that $L_{\beta}(G'_2) = \{a^n b^m c^n d^m \mid n, m \geq 1\},\ \beta \in \{mloc, mdf, mldf\}$.
Now consider the grammar

$$G''_2 = (\{a, b, c, d\}, abcd, \{(bc^+d^+, (b, d)), (bc^+, (b, d)), (bc^+, (a, c))\}).$$

Any derivation in the mode *arc* in G''_2 is as follows:

$$a^{\downarrow}bc^{\downarrow}d \Longrightarrow_{arc} a\underline{a}^{\downarrow}bc\underline{c}^{\downarrow}d \Longrightarrow_{arc} \cdots \Longrightarrow_{arc} a^{n-1}\underline{a}^{\downarrow}bc^{n-1}\underline{c}^{\downarrow}d \Longrightarrow_{arc}$$
$$a^n\underline{b}^{\downarrow}bc^n\underline{d}^{\downarrow}d \Longrightarrow_{arc} a^n b\underline{b}^{\downarrow}bc^n d\underline{d}^{\downarrow}d \Longrightarrow_{arc} \cdots \Longrightarrow_{arc} a^n b^m c^n d^m, n, m \geq 1.$$

It can be easily seen that $L_{arc}(G''_2) = L_2$. □

Lemma 3. *(marked duplication)*
$L_3 = \{wcw \mid w \in \{a, b\}^*\} \in ICC_{\alpha}(REG),\ \alpha \in \{max, mloc, mdf, mldf, arc\}$.

Proof: Consider the following grammar

$$G_3 = (\{a, b, c\}, c, (\{cx \mid x \in \{a, b\}^*\}, \{(a, a), (b, b)\})).$$

Any derivation in α, $\alpha \in \{max, mloc, mdf, mldf, arc\}$ mode of G_3 is as follows:

$$c \Longrightarrow_\alpha \underline{w_1}^\downarrow c \underline{w_1}^\downarrow \Longrightarrow_\alpha w_1\underline{w_2}^\downarrow cw_1\underline{w_2}^\downarrow \Longrightarrow_\alpha w_1w_2\underline{w_3}^\downarrow cw_1w_2\underline{w_3}^\downarrow$$
$$\Longrightarrow_\alpha w_1w_2w_3\underline{w_4}^\downarrow cw_1w_2w_3\underline{w_4}^\downarrow \Longrightarrow_\alpha \cdots \Longrightarrow_\alpha wcw,$$

where $w_i \in \{a, b\}$, $1 \leq i \leq |w|$, $w \in \{a, b\}^*$. The reader may verify that $L_\alpha(G_3) = L_3$, $\alpha \in \{mdf, mldf, arc\}$. □

Remark. When we take $c = \lambda$ in the above grammar G_3, the derivation in the *arc* mode will be

$$\alpha \Longrightarrow_{arc} \underline{w_1}^\downarrow \underline{w_1}^\downarrow \Longrightarrow_{arc} w_1\underline{w_2}^\downarrow w_1\underline{w_2}^\downarrow \Longrightarrow_{arc} w_1w_2\underline{w_3}^\downarrow w_1w_2\underline{w_3}^\downarrow$$
$$\Longrightarrow_{arc} \cdots \Longrightarrow_{arc} ww, \text{ and thus } L_{arc}(G_3) = \{ww \mid w \in \{a, b\}^*\}.$$

(This language is known non-marked duplication). This result is important from contextual grammars point of view since no other contextual grammar is capable of generating this language [17].

3.2 Computational Complexity

In this section, we shall show that all languages in the family $ICC_{mloc}(F)$, $ICC_{arc}(F)$ for $F \in \{FIN, REG\}$ are parsable in polynomial time. That is, the computational complexity of the membership problem for these languages can be solved in polynomial time. In [3], Ilie showed that the membership problem for local contextual languages with regular choice is solvable in polynomial time using logarithmic space complexity. Due to the page restriction, we just present the result here and the reader can find the proof in [3]. As absorbing context grammars are derived from local grammars, the membership problem for $ICC_{arc}(F)$ languages follows from this result. The membership problem for other families of languages $ICC_\beta(F), \beta \in \{mdf, mldf\}$ remain open.

Theorem 1. $ICC_\alpha(F) \subseteq P$, $\alpha \in \{mloc, arc\}$, $F \in \{FIN, REG\}$.

3.3 Semilinear Problem

As mentioned earlier, the family of languages generated by internal and maximal contextual grammars contains non-semilinear languages also. This question remains open for $ICC_\alpha(F)$, $\alpha \in \{mloc, mdf, mldf\}$. But, this question is solved for $ICC_{arc}(F)$ and the result is presented below.

Theorem 2. *The family of languages generated by arc grammars with F selection,* $F \in \{FIN, REG\}$, *is semilinear.*

Proof: Let L be the language generated by the *arc* grammar $G = (V, A, P)$ where P is a set of *selector-contexts* of the form

$$\{(S_1, (u_1, v_1)), (S_2, (u_2, v_2)), \ldots, (S_n, (u_n, v_n))\}.$$

Construct a regular grammar G' as follows:

$$G' = (\{S, S_{r,s} \mid 1 \leq r, s \leq n\}, V, \{S\}, P')$$

where P' consists of the following production rules:

For all $1 \leq j, f, r, s, t \leq n$, $v_f = v'_f v''_f$, $|v'_f| = \lceil \frac{|v_f|}{2} \rceil$, $|v''_f| = \lfloor \frac{|v_f|}{2} \rfloor$,

$$(1)\ \{S \rightarrow x_1 x_2 x_3 \mid w = x_1 x_2 x_3 \in A\}$$
$$(2)\ \{S \rightarrow x_1 x_2 x_3 u_j v_j \mid x_2 \in S_j,\ (S_j, (u_j, v_j)) \in P\}$$
$$(3)\ \{S \rightarrow x_1 x_2 x_3 u_j v'_j S_{j,r} \mid x_2 v_j{}' \in S_r,\ (S_r, (u_r, v_r)) \in P\}$$
$$(4)\ \{S_{r,s} \rightarrow v''_r u_s v'_s S_{s,t} \mid x_2 v'_j \ldots v'_s \in S_t,\ (S_t, (u_t, v_t)) \in P\}$$
$$(5)\ \{S_{r,s} \rightarrow v''_r u_s v_s \mid (S_s, (u_s, v_s)) \in P,\ 1 \leq r, s \leq n\}$$

Rules 1,2,5 are terminal rules which can be used to stop the derivation whenever necessary. The non-terminals in rule 3,4 will be used to simulate the derivations for further steps. It is easy to verify from the structure of P' that $L_{arc}(G)$ and $L(G')$ are letter equivalent. Since $L(G')$ is regular, $L(G)$ is also semi-linear. □

4 Concluding Remarks

In this survey report, we have seen several classes of contextual grammars and their relevances with natural languages in the view of mildly context sensitiveness. The families of languages generated by internal, maximal, depth-first and lengthwise depth-first grammars are not found to be suitable model for the syntactical description of natural languages. For maximal local grammar, the question on semilinearity is left open and for maximal depth-first and maximal lengthwise depth-first grammars, the membership problem and semilinearity are left open and so their feasibility for natural languages is still remains open. But, for absorbing right context grammars, it is shown that it satisfies all the three basic properties of mildly context sensitive formalisms and so this could be used as a syntactical description for natural languages.

Consider, the problem of parsing the maximal depth-first languages. Given this grammar and a word w, when the Turing machine initially identifies a selector in the given word, it knows that the selector contains some contexts u or v which was introduced in the previous derivation step. But this information does not help much while backtracking. So imposing the *end-marked* restriction to the maximal depth-first grammar will be of interesting and this may help to solve the membership problem in polynomial time. The end-marked depth first variant is, whenever some contexts u or v is introduced, the selector for the next derivation step should either start with one of the contexts or it should end with one of the contexts. This may help the head of the Turing matching not to move back and forth to find the selectors, rather it can start with one of the starting or end position of the selector and move in a particular direction. Hence, generating the non-context-free languages, membership problem and semilinearity issues for this end-marked depth-first variant is left as open problem.

References

[1] P. Boullier, From Contextual Grammars to Range Concatenation Grammars, *Electronic Notes in Theoretical Computer Science*, **53**, 2001.

[2] M. Daley, L. Kari, G. Gloor and Rani Siromoney, Circular Contextual Insertion/ Deletion with Applications to Biomolecular Computation, *SPIRE-'99*, 47–54, 1999.

[3] L. Ilie, On Computational complexity of contextual languages, *Theo. Comp. Sci.*, **183**(1), 33–44, 1997.

[4] L. Kari and G. Thierrin, Contextual insertions/deletions and computability, *Information and Computation*, **131**, 47–61, 1996.

[5] Kamala Krithivasan and Mutyam Madhu, Contextual P Systems, *Fundam. Inform.*, **49**(1-3), 179-189, 2002.

[6] S.N. Krishna and R. Rama, Insertion-Deletion P systems, *LNCS Proceedings of DNA7 Conference*, 360–370, 2001.

[7] S.N. Krishna, K. Lakshmanan and R. Rama, Hybrid P Systems, *Romanian Journal of Information Science and Technology*, **4**(1-2), 111–123, 2001.

[8] S.N. Krishna, K. Lakshmanan and R. Rama, On Power of P Systems with Contextual Rules, *Fundamenta informaticae*, **49**(1-3), 167–178, 2002.

[9] S.N. Krishna K. Lakshmanan and R. Rama, Tissue P Systems with Contextual and Rewriting Rules, *Workshop on Membrane Computing-2002, Curtea-de-Arges, Romania* (Gh. Păun, G. Rozenberg, A. Salomaa and C. Zandron eds.), *LNCS* - **2597**: *Membrane Computing*, 339–351, 2003.

[10] S.N. Krishna, K. Lakshmanan and R. Rama, On some classes of contextual grammars, *Intern. J. of Comp. Math.*, **80**(2), 151–164, 2003.

[11] K. Lakshmanan, S.N. Krishna, R. Rama and C. Martin-Vide, Contextual Grammars for Mildly Context Sensitive Languages, *Communicated*, 2005.

[12] M. Kudlek, C. Martin-Vide, A. Mateescu, An Infinite hierarchy of mildly context-sensitive families of languages, *TUCS Technical Report No.* **163**, 1998.

[13] K. Lakshmanan and R. Rama, On the Power of Tissue P Systems with Insertion and Deletion Rules, *Pre-proceedings of Workshop on Membrane Computing (WMC'03)*, 304 –318, 2003.

[14] K. Lakshmanan and R. Rama, The Computational Efficiency of Insertion-Deletion Tissue P Systems, to appear in *The Commemorative volume for the occasion of 75th Birth Anniversary of Dr.Rani Siromoney*, World Scientific, Vol. 666, pp. 235-245, 2006.

[15] S. Marcus, Contextual Grammars, *Rev. Roum. Pures. Appl.*, **14**, 1525–1534, 1969.

[16] S. Marcus, C. Martin-Vide and Gh. Păun, On Internal contextual grammars with maximal use of selectors, *Proc. 8th Conf. Automata and Formal Languages*, Salgotarjan, 1996.

[17] S. Marcus, C. Martin-vide and Gh. Păun, Contextual grammars as generative models of natural languages, *Computational Linguistics*, **24**(2), 245–274, 1998.

[18] C. Martin-Vide, J. Miquel-Verges and Gh. Păun, Contextual Grammars with Depth-first Derivation, *Tenth Twente Workshop on Language Technology; Algebraic Methods in Language Processing, Twente*, 225–233, 1995.

[19] Gh. Păun, Marcus Contextual Grammars, *Kluwer Academic Publishers*, 1997.

[20] Gh. Păun and X.M. Nguyen, On the Inner Contextual Grammars, *Rev. Roum. Pures. Appl.*, **25**, 641–651, 1980.

[21] A. Salomaa, Formal Languages, *Academic Press*, 1973.

K. Lakshmanan
School of Technology and Computer Science,
Tata Institute of Fundamental Research
Homi Bhabha Road, Colaba,
Mumbai - 400 005, India
email: `laksh@tifr.res.in`

Formal Language Aspects of Natural Computing– 2006, pp. 113–118

Watson-Crick *L* Systems

K.G. Subramanian

Abstract

A brief introductory survey of Watson-Crick *L* Systems that are Lindenmayer systems endowed with Watson-crick complementary features, is given in this note.

1 Introduction

Lindenmayer Systems or *L* systems, in short, were introduced by Aristid Lindenmayer in the sixties for modeling the development of multicellular organisms. The theory of *L* systems [3, 4] which constitutes a vast field of Formal language theory, gave rise to mathematically rich results in language theory besides being interesting with very many applications. A new type of rewriting, called parallel rewriting, is a basic feature of *L* systems due to the fact that these systems were intended for the purpose of modeling organisms in which development takes place in parallel and this parallelism in rewriting is in contrast to the sequential rewriting in Chomskian grammars which were motivated by linguistic considerations. Two basic references for *L* systems are [3, 4]. Also, in [5], a concise and compact account of *L* systems is provided.

DNA Computing is another area which is very relevant to Formal language theory and which has its motivation also in biology. Interest in DNA Computing has been growing very rapidly after Adleman's pioneering work on using methods of molecular biology to computational problems. The phenomenon of Watson-Crick Complementarity is basic in DNA computing. A DNA (deoxyribonucleic acid) strand is composed of bases which are denoted by A (adenine), C (cytosine), G (Guanine) and T (thymine). The DNA double helix arises by the bondage of two separate strands with the Watson-Crick complementarity (A and T are complementary; C and G are complementary) leading to the formation of such double strands. Recently complementarity has been viewed as a language-theoretic operation. Studies in this direction have enhanced the theory of formal languages.

In this survey, we provide an outline of the language-theoretic aspects of DNA Complementarity, especially in Lindenmayer systems. A simplest kind of *L* systems, called D0L systems, augmented with a specific complementarity transition are called Watson-Crick D0L Systems. The survey is intended to be a brief introduction of these systems to the uninitiated reader. The survey heavily relies on studies in [1, 2].

2 *L* Systems

An alphabet Σ is a finite set of symbols. A word w over Σ is a finite sequence of symbols from Σ. Σ^* is the set of all words over Σ. ε is the empty word with no symbols. The length of a word w, denoted by $|w|$, is the number of symbols in w.

We first provide an intuitive understanding of the main idea behind D0L-systems by considering an example:

Let us consider strings over two letters a and b . For each letter we specify a rewriting rule. The rule $a \to ab$ means that the letter a is to be replaced by the string ab, and the rule $b \to a$ means that the letter b is to be replaced by a. The rewriting process starts from a distinguished string called the axiom. Let us assume that it consists of a single letter b. In the first derivation step (the first step of rewriting) the axiom b is replaced by a using production $b \to a$. In the second step a is replaced by ab using the production $a \to ab$. The word ab consists of two letters, both of which are simultaneously replaced in the next derivation step. Thus, a is replaced by ab, b is replaced by a, and the string aba results. In a similar way (by the simultaneous replacement of all letters), the string aba yields $abaab$ which in turn yields $abaababa$, then $abaababaabaab$, and so on. The derivation is shown diagrammatically in the figure given below:

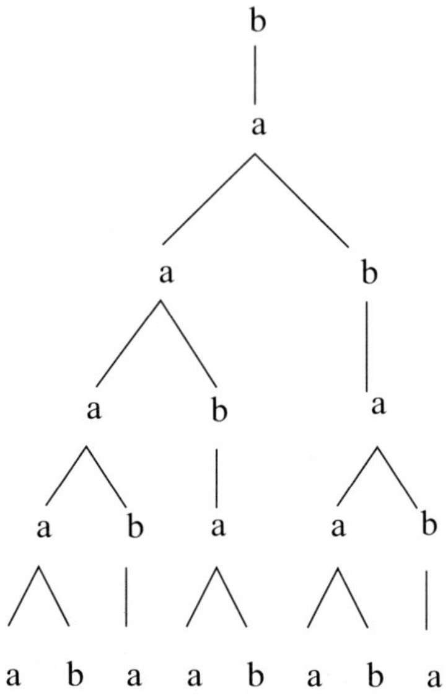

The informal illustration is now given in a formal manner. We now recall the definition of a D0L system [3, 4].

Definition 1. A 0L System is an ordered triple $G = (\Sigma, \omega, P)$ where Σ is an alphabet, ω called the axiom, is a nonempty word over Σ and P is a finite set of productions. A production (a, α) where $a \in \Sigma$ and $\alpha \in P$, is written as $a \to \alpha$. It is assumed that for any symbol $a \in \Sigma$, there is at least one production in P.

A 0L System is called a deterministic 0L system (D0L system) if there is exactly one production for every symbol in Σ.

Let $\omega_1 = a_1a_2 \dots a_m$ be a word over Σ. The word $\omega_2 = \alpha_1\alpha_2 \dots \alpha_m$ is directly derived from ω_1 if and only if $a_i \to \alpha_i$, for $i = 1, \dots, m$ is a production in P. We denote this direct derivation by $\omega_1 \Rightarrow \omega_2$. We write $\omega \Rightarrow^* \omega_n$ if $\omega_1 \Rightarrow \omega_2 \Rightarrow \omega_3 \cdots \Rightarrow \omega_n$.

The language generated by the 0L (D0L) system G is $L(G) = \{x : \omega \Rightarrow^* x, x \in \Sigma^*\}$.

Example 1. *Consider the D0L system* $G_1 = (\Sigma, \omega_1, P_1)$ *where* $\Sigma_1 = \{a, b\}$, $\omega_1 = b$, $P_1 = \{b \to a, a \to ab\}$.

Starting from axiom b, on applying the productions, we obtain the sequence of words b, a, ab, aba, abaab, abaababa, ...

Here we note that the sequence of lengths of the words is $1, 1, 2, 3, 5, 8, \ldots$ *which is the FIBONACCI Sequence!*

Remark 1. A D0L system $S = (\Sigma, \omega, P)$ can be specified by a morphism h from Σ^* to Σ^* given by $h(a) = \alpha, a \in \Sigma$ if $a \rightarrow \alpha$ is a rule in S and $h(\omega_1\omega_2) = h(\omega_1)h(\omega_2)$.

The D0L system G defines the sequence $S(G)$ of words

$$\omega_i, i \geq 0, \text{ where } \omega_{i+1} = h(\omega_i), \text{ for all } i \geq 0.$$

The language then consists of all words in $S(G)$. The length sequence is $|\omega_i|, i \geq 0$ and the growth function $f_i = |\omega_i|$.

We now mention a few of many interesting and deep mathematical results relating to D0L systems. Several equivalence problems have been investigated in the context of L systems and in particular D0L systems. The D0L equivalence problems have been of great interest giving rise to important new notions and techniques. Some of these problems are listed below.

1. The **Sequence Equivalence Problem** for D0L systems consists of deciding whether or not two given D0L systems generate the same sequence.

2. The **Language Equivalence Problem** for D0L systems consists of deciding whether or not two given D0L systems generate the same language.

3. The **Growth Equivalence Problem** for D0L systems consists of deciding whether or not two given D0L systems generate the same growth function.

All these problems are known to be decidable but the growth equivalence problem was settled first. As regards the other two, it was shown very early that decidability of one implies the decidability of the other. The decidability itself was settled only later.

We discuss here some interesting aspects of growth functions of D0L systems by considering an example.

Example 2. *Consider the D0L system* $G_2 = (\Sigma_2, \omega_2, P_2)$ *where* $\Sigma = \{a, b, c\}$, $\omega_2 = a$, $P_2 = \{a \rightarrow abcc, b \rightarrow bcc, c \rightarrow c\}$.

Starting from the axiom a, applying the production $a \rightarrow abcc$ *we have the derivation* $a \Rightarrow abcc$. *Continuing the derivation applying the productions for a, b, c and rewriting every symbol in the word abcc we obtain* $abcc \Rightarrow abccbccc$ *so that* $a \Rightarrow^* abccbccc$ *and so on. The sequence of words generated is a, abcc, abccbccc, abccbcccbccccc, ...*

We note that the sequence of lengths of these words is $1, 4, 9, 16, \ldots$ *which are SQUARES!*

The growth matrix M of a D0L system is a matrix where the ith row indicates the number of occurrences of each letter on the right of the production for each letter.

Here the Growth matrix of the system G_2 *is given by*

$$M = \begin{pmatrix} 1 & 1 & 2 \\ 0 & 1 & 2 \\ 0 & 0 & 1 \end{pmatrix}$$

where the entries 1, 1, 2 *in the first row indicate that each of the letters a, b occurs only once but c occurs twice in the right side of the rule for a; the entries* 0, 1, 2 *in the second row indicate that the letter c occurs twice, b occurs only once but c does not occur in the right side of the rule for b and so on.*

Let π *be a row vector indicating the number of occurrences of each of the letters a, b, c in the axiom of* G_2. *Here* $\pi = (1, 0, 0)$. *Let* η *be a column vector with as many rows as the number of letters in the alphabet of the D0L system. Here* η *has three rows and*

$$\eta = \begin{pmatrix} 1 \\ 1 \\ 1 \end{pmatrix}$$

Then the growth function of the system can be expressed in the form $f(i) = \pi M^i \eta$. *In fact* $f(0) = 1$, $f(1) = 4$, $f(2) = 9$, $f(3) = 16$, $f(4) = 25$, $f(5) = 36$, $f(6) = 49, \ldots$ *is the length sequence of* G_3.

The matrix representation gives a very simple method for deciding growth equivalence: Two D0L systems G and H with m and n letters respectively in their alphabets, have the same growth function if and only if the lengths of the first $m + n$ terms are the same in both the sequences.

3 Watson-Crick D0L Systems

In the formation of DNA double strands, Watson-Crick complementarity, namely, base A bonding with T and C with G, is an underlying basic feature. In other words bonding occurs if the bases in the corresponding positions are complementary. Due to the complementarity of the two strands, the information content of a double strand is the same as that of a single strand. Investigation of complementarity as a language-theoretic operation has been done in the recent past, especially in the context of D0L systems, giving rise to Watson-Crick D0L systems. Here we give an outline of these notions and a few related results [1, 2].

Let $\Sigma_{DNA} = \{A, C, G, T\}$ be called the DNA alphabet. The letter-to-letter mapping h_ω, called the Watson-Crick morphism be given by

$$h_\omega(A) = T, h_\omega(T) = A, h_\omega(C) = G, h_\omega(G) = C.$$

Complementarity can be viewed as a language-theoretic operation as follows: Whenever a "bad" string x is about to be produced by a generative process, the complementary string is taken instead of x. In other words certain undesirable conditions in the string x, trigger a transition to the complementary string. This operational complementarity has been found to be very suitable in the context of D0L systems.

Example 3. *[2] Consider a D0L system G with the alphabet* $\Sigma = \{a, t, c, g\}$, *the axiom aga and rules* $a \to a, g \to ta, c \to a, t \to g$. *Let the system have the* **trigger** $TR = \{\omega \in \Sigma^* / \omega \text{ is not in } \{a, g\}^*\}$. *The idea of the trigger is that, whenever a string not in* $\{a, g\}^*$ *is produced in the course of a derivation in the D0L system G, the complementary word must be taken. If we denote respectively by* $\Rightarrow$, *and by* $\Rightarrow_t$, *usual derivation step in a D0L system and a derivation step in the system with the trigger, then*

$$aga \Rightarrow ataa$$

must be replaced by the derivation step

$$aga \Rightarrow_t tatt,$$

since ataa is not in $\{a, g\}^*$.

Likewise tatt $\Rightarrow$ *gagg and so tatt* $\Rightarrow_t$ *gagg as gagg is in* $\{a, g\}^*$.

Thus a derivation in the G is as follows:

$$aga \Rightarrow_t tatt \Rightarrow_t gagg \Rightarrow_t attatat \Rightarrow_t aggagag \Rightarrow_t \cdots.$$

A trigger TR is **sound** if h_ω is not in TR whenever ω is in TR.

The trigger in the Example 3 is not sound.

We can now define a DNA-like alphabet $\Sigma = \{a_1, a_2, \ldots, a_n, \overline{a}_1, \overline{a}_2, \ldots, \overline{a}_n\}$ generalizing the notion of Complementarity. The letters a_i and $\overline{a}_i$ are called complementary to each other. The letter-to-letter mapping $h_\omega(x)$ given by

$$h_\omega(a_i) = \overline{a}_i \text{ and } h_\omega(\overline{a}_i) = a_i, i = 1, \ldots, n.$$

$h_\omega(x)$ is called the Watson-Crick morphism. In the case of DNA alphabet, $a_1 = A$, $a_2 = G$, $\overline{a}_1 = T$, $\overline{a}_2 = C$. The nucleotides A and G are purines; T and C are pyrimidines. Extending this terminology, the non-barred letters a_i are called purines and the barred letters $\overline{a}_i$ are called pyrimidines.

The collection PYR consists of words over Σ where the number of occurrences of pyrimidines exceeds that of purines. *PYR* is taken as the class of bad words frequently.

Definition 2. [2] A Watson-Crick D0L system is $G = (\Sigma, g, \omega_0)$ where

$$\Sigma = \{a_1, a_2, \ldots, a_n, \overline{a}_1, \overline{a}_2, \ldots, \overline{a}_n\},$$

$g : \Sigma^* \rightarrow \Sigma^*$ is a morphism and ω_0, the axiom, is a nonempty word in the set PUR, the complement of *PYR*.

The sequence $S(G)$ defined by G consists of words $\omega_0, \omega_1, \ldots$ where for $i \geq 0$, $\omega_{i+1} = g(\omega_1)$ if $g(\omega_i) \in$ PUR and $= h_\omega(g(\omega_i)) \in PYR$.

We say that $S(G)$ is defined using a D0L system augmented with the trigger *PYR*. The language $L(G)$ generated by G consists of all words in the sequence $S(G)$. We note that the trigger *PYR* is sound.

We can consider sequence, language, and growth equivalence problems for Watson-Crick D0L systems as done for D0L systems. Equivalence problems are largely open for these systems. There are Watson-Crick D0L systems for which growth fluctuates between linear growth and exponential growth and such growth type is strange from the point of view of D0L growth functions.

Many such interesting new notions have been investigated in the literature in the context of Watson-Crick D0L systems.

Conclusion

Lindenmayer systems and in particular, D0L systems have given rise to not only rich theoretical results but have also been applied in many application areas, such as computer graphics. The purpose of this introductory survey is to draw the attention of the interested reader to the new notion of Watson-Crick D0L systems that are based on concepts in DNA computing. Many theoretical questions in this area remain awaiting solution Also possible applications of these systems could be examined. The author wishes to acknowledge the fact that the material presented here are drawn from [1, 2].

References

[1] J. Honkala and A. Salomaa, Watson-Crick D0L Systems with Regular Triggers, Theoret. Comput. Sci., 259 (2001) 689-698.

[2] V. Mihalache and A. Salomaa, Language-theoretic aspects of DNA complementarity, Theoret. Comput. Sci., 250 (2001) 163-178.

[3] G. Rozenberg and A. Salomaa, The Mathematical Theory of *L* Systems, Academic Press, New York, London, 1980.

[4] G. Rozenberg and A. Salomaa (eds), Handbook of Formal Languages, Vol. 1, Springer Verlag, Berlin, Heidelberg, Academic Press, New York, 1997.

[5] A. Salomaa, Formal Languages, New York, 1973.

K.G. Subramanian
Department of Mathematics
Madras Christian College
Chennai 600 059

Formal Language Aspects of Natural Computing– 2006, pp. 119–142

Network of Evolutionary Processors, Past, Present and Future

Ashish Choudhary

Abstract

A network of evolutionary processors (NEP) consists of several language processors which are located in the nodes of a virtual graph and are able to perform only one type of point mutations (insertion, deletion or substitution of symbols) on the words found in that node, according to some predefined rules. Each node is associated with an input filter and an output filter, defined by some regular language. After applying in parallel a point mutation to all the words existing in every node, the new words which are able to pass the output filter of the respective node navigate simultaneously through the network and enter those nodes whose input filter they are able to pass. We survey several variants of NEP and their computational power. Some open problems and direction of further research are finally pointed out.

1 Introduction

In the later half of the last century, electronic computers have played a big role in helping mankind to solve difficult problems and in day to day life. Due to the tremendous changes the electronic computers have brought about, the world is looking upon the electronic computers as the ultimate computing machines. However, some computer theorists are looking beyond the conventional computational models and are examining completely new computational models. A natural question arises - why to look for new models of computation, when electronic computers are dominating the world and seem to be very good? There are two fundamental reasons behind this.

One of them is that there are several problems, termed as intractable problems, which require large amount of running time even on supercomputers. These problems take huge amount of processing time and the only way of solving these problems is through exhaustive search in the sample solution space of the problem. This exhaustive search done in any electronic computer, which is deterministic requires exponential running time.

The other fundamental reason is, our technology is growing faster and this growth is a fleeting one. There is a fundamental limit on how fast an electron can travel through a conductor [28] and hence we have a restriction on the speed of communication between communicating parts inside a machine.

To overcome these shortcomings, many new paradigms of computing have been introduced recently. They include DNA computing [36], quantum computing [36] and membrane computing [1]. All these new computational models have been proved to be computationally complete by proving their equivalence to Turing machines, see [36] and their references. We can solve NP-complete problems in polynomial time using these models because of the massive parallelism involved in these models.

The origin of networks of evolutionary processors (NEPs for short) is twofold. In [15], the authors have considered a computing model inspired by the evolution of cell populations, which might model some properties of evolving cell communities at the syntactic level. Cells are represented by words which encode their DNA sequences. Informally, at any moment of time, the evolutionary system is described by a collection of words, where each word represents one cell. Cells belong to species and their community evolves according to mutations and division which are defined by operations on words. Only those cells are accepted as surviving (correct) ones which are represented by a word in a given set of words, called the genotype space of species. This feature parallels with the natural process of evolution.

On the other hand, a well known architecture for parallel and distributed symbolic processing, related to the Connection Machine [26] as well as Logic Flow paradigm [19], consists of several processors, each of them being placed in a node of virtual complete graph, which are able to handle data associated with the respective node. Each node processor acts on the local data in accordance with some predefined rules and then the local data becomes a mobile agent which can navigate in the network following a given protocol. Only the data which can pass a filtering process can be communicated among the processors. The filtering process may require to satisfy some conditions imposed by the sending processor, by the receiving processor or by both of them. All the nodes send simultaneously their data and the receiving nodes also handle simultaneously all the arriving data, according to some strategies.

Starting from the premise that data can be given in the form of strings, [16] introduces a concept called network of parallel language processors in the aim of investigating this concept in terms of formal languages and grammars. Networks of language processors are closely related to grammar systems, more specifically to parallel communicating grammar systems [13]. The main idea is that one can place a language generating device (grammar, Lindenmayer system, etc.) in any node of an underlying graph which rewrites the words existing in the node, then the words are communicated to other nodes. Words can be successfully communicated if they pass some output and input filter.

In [3], the authors have modified the above concept, inspired from cell biology. Each processor placed in a node is a very simple processor, an evolutionary processor. An evolutionary processor is a processor which is able to perform very simple operations, namely point mutations in a DNA sequence (insertion, deletion or substitution of a pair of nucleotides). More generally, each node may be viewed as a cell having genetic information encoded in DNA sequences which may evolve by local evolutionary events, that is point mutations. Each node is specialized just for one of these evolutionary operations. Furthermore, the data in each node is organized in the form of multisets of words (each word appearing in an arbitrarily large number of copies) and all the copies are processed in parallel such that all possible events that can take place do actually take place. The computational power of these devices are investigated in [4, 3]. Furthermore, filters, based on membership and random-context conditions, used in [3] are generalized in some versions defined in [33]. More precisely, the new filters are based on different types of random-context conditions. Thus, in the new model, rules are applied in different manner in each node. Also, different nodes have different types of filters. Due to these reasons, the new model is called hybrid networks of evolutionary processors (HNEP in short) [33]. HNEPs are shown to be universally complete [14]. The size descriptional complexity measures of HNEPs as generating devices are studied in [2].

On the other hand, focus is also there on the way networks of evolutionary processors are used in solving membership decision problems. For this reason, accepting hybrid networks of evolutionary processors (AHNEPs) were defined in [32]. Such a network contains two distinguished nodes: the input and the output node. At the beginning of the computation the only node that contains local data is the input node and this local data consists of the string that we want to decide whether it is a member of a given language or not. Then the appliance of evolutionary rules and communication are performed alternatively, until the output node receives some data or until all the nodes process in a computational step exactly the same data that it processed in the precedent step. In the first case the computation is successful and the input string is accepted, but in the second one and also in the case of an infinite computation, the input string is rejected. The computational power of AHNEP are investigated in [32, 30].

In addition to NEPs, HNEPs and AHNEPs some other variants, properties and application of network of evolutionary processors are studied in [34, 31, 17, 18].

In the present paper, we survey some results regarding all aforementioned models from the point of view of their computational power. We also introduce some new variants of NEP, HNEP and AHNEP and state their properties.

The paper is organized as follows: In the next section, we give some pre-requisites from formal language theory and recall the definition of NEP from [3]. We also state the power of NEP as given in [4, 3]. We then give the definition of HNEP in Section 3 and state their properties. This is followed by the definition of AHNEP in Section 4. In Section 5 we state some of the application of network of evolutionary processors. We then introduce several other variants of NEP, HNEP and AHNEP in Section 6. The paper ends with a brief conclusion and directions for further research.

2 Network of Evolutionary Processors

Here we give some pre-requisites from formal language theory. We also recall the definition of NEP and their properties from [4, 3].

The set of all strings over V is denoted by V^* and the empty string is denoted by ε. A multiset over a set X is a mapping $M : X \to \mathbf{N} \cup\{\infty\}$. The number $M(x)$ expresses the number of copies of $x \in X$ in the multiset M. When $M(x) = \infty$, then x appears arbitrarily many times in M. The set $supp(M)$ is the support of M, i.e., $supp(M) = \{x \in X | M(x) > 0\}$.

A *network of evolutionary processors* (NEP for short) of size n is a construct

$$\Gamma = (V, N_1, N_2, \ldots, N_n, G),$$

where V is an alphabet and for each $1 \leq i \leq n$, $N_i = (M_i, A_i, PI_i, PO_i)$ is the i–th evolutionary node processor of the network. The parameters of every processor are:

- M_i is a finite set of evolution rules of one of the following forms only
  - $a \to b$, $a, b \in V$ (substitution rules),
  - $a \to \varepsilon$, $a \in V$ (deletion rules),
  - $\varepsilon \to a$, $a \in V$ (insertion rules).

  More clearly, the set of evolution rules of any processor contains either substitution or deletion or insertion rules.

- A_i is a finite set of strings over V. The set A_i is the set of initial strings in the i–th node. We consider that each string appearing in any node at any step has an arbitrarily large number of copies in that node.

- PI_i and PO_i are subsets of V^* representing the input and output filter respectively. These filters are defined by membership condition, namely a string $w \in V^*$ can pass the input filter (the output filter) if $w \in PI_i$ ($w \in PO_i$).

Finally $G = (\{N_1, N_2, \ldots, N_n\}, E)$ is an undirected graph called the *underlying graph* of the network. By a configuration (state) of an NEPS as above we mean an n–tuple $C = (L_1, L_2, \ldots, L_n)$, with $L_i \subseteq V^*$ for all $1 \leq i \leq n$. A configuration represents the sets of strings (each string appears in an arbitrarily large number of copies) which are present in any node at a given moment. The initial configuration of the network is $C_0 = (A_1, A_2, \ldots, A_n)$. A configuration can either change by an evolutionary step or by a communicating step. When changing by an evolutionary step, each component L_i of the configuration is changed in accordance with the rewriting rules associated with the node i.

Formally, we say that the configuration $C_1 = (L_1, L_2, \ldots, L_n)$ directly changes into the configuration $C_2 = (L'_1, L'_2, \ldots, L'_n)$ by an evolutionary step, written as $C_1 \Rightarrow C_2$ if L'_i is the set of strings obtained by applying the rules of M_i to the strings in L_i as follows:

(i) If the same substitution or deletion rule can replace different occurrences of the same symbol within a string, then all these occurrences must be replaced within different copies of that string. The result is a multiset in which every string that can be obtained by using these rules appears in an arbitrarily large number of copies.

(ii) An insertion rule is applied at any position in a string. Again, the result is a multiset in which every string that can be obtained by application of an insertion rule to any arbitrary position in an existing string appears in an arbitrarily large number of copies.

(iii) If more than one rule, no matter of its type can be applied to a string, then all of them must be used for different copies of that string.

In other words, since an arbitrarily large number of copies of each string is available in every node, after an evolutionary step in each node one gets an arbitrarily large number of copies of any strings which can be obtained by using any rule from the set of evolution rules associated with that node. By definition, if L_i is empty for some $1 \leq i \leq n$, then L'_i is also empty.

When changing by communication step, each node processor N_i sends all copies of the strings it has which are able to pass its output filter to all the node processors connected to N_i and receives all copies of the strings sent by any node processor connected with N_i providing they can pass its input filter. Formally we say that the configuration $C_1 = (L_1, L_2, \ldots, L_n)$ directly changes into the configuration $C_2 = (L'_1, L'_2, \ldots, L'_n)$ by a communication step, written as $C_1 \vdash C_2$ if

$$L'_i = L_i - \{w | w \in L_i \cap PO_i\} \cup \bigcup_{\{N_i, N_j\} \in E} \{x | x \in L_j \cap PO_j \cap PI_i\}.$$

Let $\Gamma = (V, N_1, N_2, \ldots, N_n)$ be a NEP as defined above. By a computation in Γ we mean a sequence of configurations $C_0, C_1, C_2, \ldots$, where C_0 is the initial configuration, $C_{2i} \Rightarrow C_{2i+1}$ and $C_{2i+1} \vdash C_{2i+2}$ for all $i \geq 0$.

If the sequence is finite, then we have a finite computation. The result of any finite or infinite computation is a language which is collected in a designated node called the output (master) node of the network. If one considers the output node of the network as being the node k, and if $C_0, C_1, \ldots$ is a computation, then all strings existing in the node k at some step t - the kth component of C_t belongs to the language generated by the network. We will denote this language by $L_k(\Gamma)$.

The time complexity of computing a finite set of strings Z is the minimal number of steps t in a computation $C_0, C_1, \ldots, C_t \ldots$ such that Z is a subset of the kth component of C_t.

The following theorems are taken from [3].

Theorem 1 ([3]). *Each recursively enumerable language can be generated by a complete NEP of size 5.*

Theorem 2 ([3]). *Each recursively enumerable language can be generated by a star NEP of size 5.*

Theorem 3 ([3]). *Each recursively enumerable language can be generated by a ring NEP of size 6.*

The above theorems shows the power of NEP as a language generating device. The following theorems taken from [4, 3] shows the power of NEP as a computational device.

Theorem 4 ([4]). *The Bounded Post Correspondence Problem (BPCP) can be solved by a NEP in size and time linearly bounded by the product of K and the length of the two Post lists.*

BPCP is a variant of much celebrated computer science problem called the Post Correspondence Problem (PCP) known to be unsolvable [20]. An instance of the PCP consists of an alphabet V and two lists of strings over V

$$u = (u_1, u_2, \ldots, u_n) \text{ and } v = (v_1, v_2, \ldots, v_n).$$

The problem asks whether or not a sequence $i_1, i_2, \ldots, i_k$ of positive integers exists, each between 1 and n, such that

$$u_{i_1} u_{i_2} \ldots u_{i_k} = v_{i_1} v_{i_2} \ldots v_{i_k}.$$

The problem is undecidable when no upper bound is given for k and NP-complete when k is bounded by a constant $K \leq n$.

Theorem 5 ([3]). *The "3-colorability problem" can be solved in $O(n + m)$ time by a complete simple NEP of size $7m + 2$, where n is the number of vertices and m is the number of edges of the input graph.*

A simple NEP is a slight variation of NEP, where all the parameters are same as that of NEP except the filters. In a NEP the filters are based on regular sets where as in a simple NEP the filters are based on random context conditions [3].

3 Hybrid Network of Evolutionary Processors

Here we recall the definition of hybrid network of evolutionary processors (HNEP in short) from [33]. We also state the properties of HNEP from [14, 2, 41].

For each nonempty word x, $alph(x)$ is the minimal alphabet W such that $x \in W^*$. Let V be an alphabet. We say that a rule $a \to b$, with $a, b \in V \cup \{\varepsilon\}$ is a *substitution rule* if both a and b are different from ε; it is a *deletion rule* if $a \neq \varepsilon$ and $b = \varepsilon$; it is an *insertion rule* if $a = \varepsilon$ and $b \neq \varepsilon$. The set of all substitution, deletion and insertion rules over an alphabet V are denoted by Sub_V, Del_V and Ins_V respectively. Given a rule σ as above and a word $w \in V^*$, we define the following *actions* of σ on w:

- If $\sigma \equiv a \to b \in Sub_V$, then $\sigma^*(w) = \left\{ \begin{array}{l} \{ubv : \exists u, v \in V^*(w = uav)\}, \\ \{w\}, \text{otherwise.} \end{array} \right\}$

- If $\sigma \equiv a \to \varepsilon \in Del_V$, then $\sigma^*(w) = \left\{ \begin{array}{l} \{uv : \exists u, v \in V^*(w = uav)\}, \\ \{w\}, \text{otherwise.} \end{array} \right\}$

$$\sigma^r(w) = \left\{ \begin{array}{l} \{u : w = ua\}, \\ \{w\}, \text{otherwise.} \end{array} \right\} \qquad \sigma^l(w) = \left\{ \begin{array}{l} \{v : w = av\}, \\ \{w\}, \text{otherwise.} \end{array} \right\}$$

- If $\sigma \equiv \varepsilon \to a \in Ins_V$, then

$$\sigma^*(w) = \{uav : \exists u, v \in V^*(w = uv)\}, \sigma^r(w) = \{wa\}, \sigma^l(w) = \{aw\}.$$

Symbol $\alpha \in \{*, l, r\}$ expresses the way of applying a rule to a word, namely at any position ($\alpha = *$), in the left ($\alpha = l$) or in the right ($\alpha = r$) end of the word, respectively. For every rule σ, action $\alpha \in \{*, l, r\}$ and $L \subseteq V^*$, we define the α-action on the word w and the language L by:

$$M^\alpha(w) = \bigcup_{\sigma \in M} \sigma^\alpha(w) \quad \text{and} \quad M^\alpha(L) = \bigcup_{w \in L} M^\alpha(w),$$

respectively. In what follows, we shall refer to the rewriting operations defined above as *evolutionary operation* since they may be viewed as linguistic formulations of local gene mutations. For two disjoint and nonempty subsets P and F of an alphabet V and a word over V, we define the predicates

$$\begin{array}{lll} \varphi^{(1)}(w; P, F) \equiv P \subseteq alph(w) & \wedge & F \cap alph(w) = \emptyset \\ \varphi^{(2)}(w; P, F) \equiv alph(w) \subseteq P & & \\ \varphi^{(3)}(w; P, F) \equiv P \subseteq alph(w) & \wedge & F \not\subseteq alph(w) \\ \varphi^{(4)}(w; P, F) \equiv alph(w) \cap P \neq \emptyset & \wedge & F \cap alph(w) = \emptyset. \end{array}$$

The construction of these predicates is based on *random-context conditions* defined by the two sets P (*permitting contexts*) and F (*forbidding contexts*).

For every language $L \subseteq V^*$ and $\beta \in \{(1), (2), (3), (4)\}$, we define:

$$\varphi^\beta(L, P, F) = \{w \in L | \varphi^\beta(w; P, F)\}.$$

An *evolutionary processor over* V is a tuple (M, PI, FI, PO, FO), where:

- Either ($M \subseteq Sub_V$) or ($M \subseteq Del_V$) or ($M \subseteq Ins_V$). The set M represents the set of evolutionary rules of the processor. As one can see, a processor is "specialized" in one evolutionary operation, only.

- PI, $FI \subseteq V$ are the *input* permitting/forbidding contexts of the processor, while PO, $FO \subseteq V$ are the *output* permitting/forbidding contexts of the processor.

We denote the set of evolutionary processors over V by EP_V.

A *hybrid network of evolutionary processors* (HNEP for short) is a 7-tuple

$$\Gamma = (V, G, N, C_0, \alpha, \beta, i_o), \text{ where :}$$

- V is an alphabet
- $G = (X_G, E_G)$ is a undirected graph, with the set of vertices X_G and the set of edges E_G. G is called the *underlying graph* of the network.
- $\mathcal{N} : X_G \rightarrow EP_V$ is a mapping which associates with each node $x \in X_G$ the evolutionary processor $\mathcal{N}(x) = (M_x, PI_x, FI_x, PO_x, FO_x)$.
- $C_0 : X_G \rightarrow V^*$ is a mapping which identifies the initial configuration of the network. It associates a finite set of words with each node of the graph G.
- $\alpha : X_G \rightarrow \{*, l, r\}; \alpha(x)$ gives the action mode of the rules of node x on the words existing in that node.
- $\beta : X_G \rightarrow \{(1), (2), (3), (4)\}$ defines the type of the *input/output filters* of a node. More precisely, for every node, $x \in X_G$, the following filters are defined:

$$\text{input filter} \ : \rho_x(.) = \varphi^{\beta(x)}(.; PI_x, FI_x),$$

$$\text{output filter} \ : \tau_x(.) = \varphi^{\beta(x)}(.; PO_x, FO_x).$$

  That is, $\rho_x(w)$ (resp. τ_x) indicates whether or not the word w can pass the input (resp. output) filter of x. More generally, $\rho_x(L)$ (resp. $\tau_x(L)$) is the set of words of L that can pass the input (resp. output) filter of x.
- $i_o \in X_G$ is the *output node* of the HNEP.

We say that $card(X_G)$ is the size of Γ. If $\alpha(x) = \alpha(y)$ and $\beta(x) = \beta(y)$ for any pair of nodes $x, y \in X_G$, then the network is said to be *homogeneous*. Similar to NEP, a HNEP is called a *star, ring* or *grid* HNEP if its underlying graph is star, ring or grid respectively.

A *configuration* of a HNEP Γ as above is a mapping $C : X_G \rightarrow V^*$ which associates a set of words with every node of the graph. A configuration may be understood as the set of words which are present in any node at a given moment. A configuration can change either by an *evolutionary step* or by a *communication step*. When changing by an evolutionary step, each component $C(x)$ of the configuration C is changed in accordance with the set of evolutionary rules M_x associated with the node x and the way of applying these rules $\alpha(x)$. Formally, we say that the configuration C' is obtained in *one evolutionary step* from the configuration C, written as $C \Longrightarrow C'$, iff

$$C'(x) = M_x^{\alpha(x)}(C(x)) \text{ for all } x \in X_G.$$

When changing by a communication step, each node processor $x \in X_G$ sends one copy of each word it has, which is able to pass the output filter of x, to all the node processors connected to x and receives all the words sent by any node processor connected with x providing that they can pass its input filter.

Formally, we say that the configuration C' is obtained in *one communication step* from the configuration C, written as $C \vdash C'$, iff

$$C'(x) = (C(x) - \tau_x(C(x))) \cup \bigcup_{\{x,y\} \in E_G} (\tau_y(C(y)) \cap \rho_x(C(y))) \text{ for all } x \in X_G.$$

Let Γ be an HNEP, the computation of Γ on the input word $w \in V^*$ is a sequence of configurations $C_0^{(w)}, C_1^{(w)}, C_2^{(w)}, \ldots,$ where $C_0^{(w)}$ is the initial configuration of Γ, $C_{2i} \Longrightarrow C_{2i+1}$ and

$C_{2i+1} \vdash C_{2i+2}$, for all $i \geq 0$. If the sequence is finite, we have a finite computation. If we use HNEPs as language generating devices, then the generated language is the set of all words which appear in the output node under a finite computation in the network. Formally, the *language* generated by Γ is

$$L(\Gamma) = \bigcup_{s \geq 0} C_s(i_o).$$

3.1 Computational Completeness of HNEP

The following theorems are taken from [14] which shows that HNEP are computationally complete.

Theorem 6 ([14]). *Let $G = (N, T, P, S)$ be a phrase-structure grammar in the Geffert normal form, namely with $N = \{S, A, B, C\}$ and rules of the form $S \rightarrow x$, $x \in (N \cup T)^+$, and $ABC \rightarrow \varepsilon$. G can be generated by a complete HNEP.*

The number of nodes required in the above network is given by

$$s = 28 + 2m + 2\sum_{r=1}^{m} p_r + 4.card(T).$$

where m is the number of production rules and T is the set of terminals. The above network can be modified slightly so that the above theorem holds for star HNEP also [14].

Theorem 7 ([14]). *1. Any recursively enumerable language over an alphabet V can be generated by a complete HNEP of size* $27 + 3.\ card(V)$.

2. Any recursively enumerable language over an alphabet V can be generated by a star HNEP of size $28 + 3.\ card(V)$.

Theorem 8 ([14]). *1. Every language generated by an HNEP of size one is regular.*

2. There exist context-free non regular (even non-linear) languages which can be generated by complete HNEPs of size 2.

3. There exist context-sensitive non-context-free languages which can be generated by complete HNEPs of size 4.

4. There exist non-recursive languages which can be generated by complete HNEPs of size 27.

5. There exist non-recursive languages which can be generated by star HNEPs of size 28.

The above theorems show that HNEP are computationally complete.

3.2 Size Complexity of Complete HNEPs

The size descriptional complexity measures of HNEPs as language generating devices are studied in [2]. Let L be a language generated by a complete HNEP. We define

$$size(L) = min\ \{size(\Gamma) | L = L(\Gamma)\}.$$

The following theorems are taken from [2].

Theorem 9 ([2]). *Any regular language L over an alphabet with n symbols can be generated by a complete HNEP of size* $2n + 3$. *Hence,* $size(L) \leq 2.card(alph(L)) + 3$.

Theorem 10 ([2]). *For any regular language L,*

$$size(L) \leq min(2.card(alph(L)) + 3, card(alph(L)) + 5).$$

Theorem 11 ([2]). *For any regular language L accepted by an n-state minimal DFA,* $size(L) \leq 2n + 2$.

Theorem 12 ([2]). *Given a regular language L one can algorithmically decide whether or not* $size(L) = 1$.

Theorem 13 ([2]). *One cannot algorithmically decide whether the size of a context free language equals 1.*

We finish this section with two open problems given in [14]:

1. Is it possible to generate an arbitrary recursively enumerable language over an alphabet V with a complete/star HNEP of size smaller than $27 + 3.card(V)$.

2. What is the size of the smallest complete/star/ring HNEP which generates a non-context-sensitive or non-recursive language?

3.3 Solving NP-complete problems using HNEP

Consider the following three problems:

1. *The maximal independent set:* Given an undirected graph $G = (V, E)$, where V is the finite set of vertices and E is the set of edges given as a family of sets of two vertices, find the cardinality of a maximal subset (with respect to inclusion) of V which does not contain both vertices connected by any edge in E.

2. *The vertex cover problem:* Given an undirected graph find the cardinality of the minimal set of vertices such that each edge has at least one of its extremes in this set.

3. *The satisfiability problem:* For a given set P of Boolean variables and a finite set U of clauses over P, does a truth assignment for the variables of P exist satisfying all the clauses of U?

For detailed formulations and discussions about their solutions, the reader is referred to [20].

These problems can be viewed as special cases of the following algorithmic problem, called the Common Algorithmic Problem (CAP) discussed in [25]: let S be a finite set and F be a non-empty family of subsets of S. Find the cardinality of a maximal subset of S which does not include any set belonging to F. The sets in F are called *forbidden sets*. We say that (F, S) is an $(card(S), card(F))$-instance of CAP.

Let us see how the three problems mentioned above can be obtained as special cases of CAP. For the first problem, we just take $S = V$ and $F = E$. The second problem is obtained by letting $S = V$ and F contains all sets $o(x) = \{x\} \cup \{y \in X | \{x, y\} \in E\}$. The cardinality one looks for is the difference between the cardinality of S and the solution of CAP. The third problem is obtained by letting $S = P \cup P'$, where $P' = \{p' | p \in P\}$ and $F = \{F(C) | C \in U\}$, where each set $F(C)$ associated with the clause C is defined by

$$F(C) = \{p' | p \text{ appears in } C\} \cup \{p | \neg p \text{ appears in } C\}.$$

From this it follows that the given instance of the satisfiability problem has a solution if and only if the solution of the constructed instance of the CAP is exactly the cardinality of P.

Theorem 14 ([33]). *Let* $(S = \{a_1, a_2, \ldots, a_n\}, F = \{F_1, F_2, \ldots, F_m\})$ *be an* (n, m)*-instance of the CAP. It can be solved by a complete HNEP of size* $m + 2n + 2$ *in* $O(m + n)$ *time.*

As far as the time and memory resources that the above HNEP uses, the total number of symbols is $2m + 5n + 4$, see [33] and total number of rules are

$$mn + m + 5n + 2 + \sum_{i=1}^{m} card(F_i) \in \Theta(mn).$$

The same problem can be solved in a more economic way, regarding especially the number of rules, namely

Theorem 15 ([33]). *Any instance of the CAP can be solved by a complete HNEP of size m+n+1 in O(m+n) time.*

The above theorems show the power of HNEPs as a computational device.

4 Accepting Hybrid Network of Evolutionary Processors

Both network of evolutionary processors and hybrid network of evolutionary processors are language generating devices. Both the models generate all the strings belonging to a language. On the other hand, focus is also there on the way networks of evolutionary processors are used in solving membership decision problems. For this reason, accepting hybrid networks of evolutionary processors (AHNEPs) are defined in [32]. Such a network contains two distinguished nodes: the input and the output node. At the beginning of the computation the only node that contains local data is the input node and this local data consists of the string that we want to decide whether it is a member of a given language or not. Then the appliance of evolutionary rules and communication are performed alternatively, until the output node receives some data or until all the nodes process in a computational step exactly the same data that it processed in the precedent step. In the first case the computation is successful and the input string is accepted, but in the second one and also in the case of an infinite computation, the input string is rejected.

An *accepting hybrid network of evolutionary processors* (AHNEP for short) is a 7-tuple

$$\Gamma = (V, U, G, N, \beta, x_I, x_O), \text{ where :}$$

- V and U are the input and network alphabet respectively, with $V \subseteq U$.
- $G = (X_G, E_G)$ is an undirected graph with the set of vertices X_G and the set of edges E_G. G is called the *underlying graph* of the network.
- $N : X_G \to EPSP_U$ is a mapping which associates with each node $x \in X_G$ an evolutionary $N(x) = (M_x, PI_x, FI_x, PO_x, FO_x)$.

- $\beta : X_G \rightarrow \{(1), (2), (3), (4)\}$ defines the type of the *input/output filters* of a node as in HNEP.

- x_I and $x_O \in X_G$ is the *input node* and the *output node*, respectively of the AHNEPSP.

A configuration (state) of a AHNEP as above is an n-tuple $C = (L_1, L_2, \ldots L_n)$, with $L_i \subseteq V^*$ for all $1 \leq i \leq n$. A configuration represents the sets of strings (each string appears in an arbitrarily large number of copies) which are present in any node at a given moment. Given a word $w \in V^*$, the initial configuration of Γ on w is defined by $C_0^{(w)}(x_I) = w$ and $C_0^{(w)}(x) = \emptyset$ for all $x \in X_G - \{x_I\}$. A configuration can either change by an evolutionary step or by a communicating step. When changing by an evolutionary step, each component L_i of the configuration is changed in accordance with the rules M_x associated with the node i.

When changing by communication step, each node processor N_i sends all copies of the strings it has which are able to pass its output filter to all the node processors connected to N_i and receives all copies of the strings sent by any node processor connected with N_i providing they can pass its input filter. Formally, the configuration C' is obtained in *one communication step* from configuration C, written as $C \vdash C'$, iff

$$C'(x) = (C(x) - \tau_x(C(x))) \cup \bigcup_{\{x,y\} \in E} (\tau_y(C(y)) \cap \rho_x(C(y)))$$

for all $x \in G$.

Here E is the set of edges of the underlying graph of the network. Let Γ be an AHNEP, the computation of Γ on the input word $w \in V^*$ is a sequence of configurations $C_0^{(w)}, c_1^{(w)}, c_2^{(w)}, \ldots$, where $C_0^{(w)}$ is the initial configuration of Γ defined by $C_0^{(w)}(x_I) = w$ and $C_0^{(w)}(x) = \emptyset$ for all $x \in X_G$, $x \neq x_I$, $C_{2i}^{(w)} \Rightarrow C_{2i+1}^{(w)}$ and $C_{2i+1}^{(w)} \models C_{2i+2}^{(w)}$, for all $i \geq 0$. By the previous definitions, each configuration $C_i^{(w)}$ is uniquely determined by the configuration $C_{i-1}^{(w)}$. Otherwise stated, each computation in an AHNEP is deterministic. A computation as above halts if one of the following two conditions holds:

(i) There exists a configuration in which the set of words existing in the output node x_O is non-empty. In this case, the computation is said to be an *accepting computation*

(ii) There exist two consecutive identical configurations.

In the aforementioned cases the computation is said to be finite. The language accepted by Γ is

$$L(\Gamma) = \{w \in V^* | \text{ the computation of } \Gamma \text{ on } w \text{ is an accepting one } \}.$$

4.1 Complexity Classes Based on AHNEPs

Here we recall the definition of the complexity classes based on AHNEPs from [32]. The reader is referred to [22, 23] for the classical time and space complexity classes defined on the standard computing model of Turing machine.

Consider a AHNEP Γ and the language L accepted by Γ. The *time complexity* of the accepting computation $C_0^{(x)}, C_1^{(x)}, C_2^{(x)}, \ldots, C_m^{(x)}$ of Γ on $x \in L$ is denoted by $Time_\Gamma(x)$ and equals m. The time complexity of Γ is the partial function from $\mathbf{N}$ to $\mathbf{N}$,

$$Time_\Gamma(n) = \max \{Time_\Gamma(x) | x \in L(\Gamma), |x| = n\}.$$

For a function $f : \mathbf{N} \to \mathbf{N}$ we define

$$\mathbf{Time}_{AHNEP}(f(n)) = \{L \mid \text{there exists an AHNEP } \Gamma \text{ and } n_0 \text{ such that}$$

$$L = L(\Gamma) \text{ and } \forall n \geq n_0(Time_\Gamma(n) \leq f(n))\}$$

Moreover, we write $\mathbf{PTime}_{AHNEP} = \bigcup_{k \geq 0} \mathbf{Time}_{AHNEP}(n^k)$.

The following results are taken from [32] which establishes a strong connection between the complexity classes defined on Turing machines and those defined on AHNEPs.

Proposition 1 ([32]). *For any nondeterministic Turing machine M recognizing a language L there exists an AHNEP Γ accepting the same language L. Moreover if M works within time $f(n)$ then $Time_\Gamma(n) \in O(f(n))$.*

Theorem 16 ([32]). $\mathbf{NP} = \mathbf{PTime}_{AHNEP}$.

4.2 Solving NP-complete Problems using AHNEPs

In [30] it is shown how to solve 3CNF-SAT and Hamiltonian path problem (HPP) in *linear* time using AHNEPs.

The SAT problem is as follows: Given a formula E in 3CNF with n variables and m clauses, does there exists an argument of the n boolean variables such that the m clauses of 3 different variable each are all satisfied. The problem is NP-complete [20]. In [30] it is shown that the above problem can be solved by using a complete AHNEP of size $2n + 2$ in $2n + 6m$ steps.

The HPP problem is to decide whether or not a given directed graph has a Hamiltonian path. A Hamiltonian path in a directed graph is a path which contains all the vertices exactly once. The HPP problem is NP-complete [20]. In [30] it is shown how to solve the above problem using a complete AHNEP of size $2n+1$ in $4n$ steps where n is the number of vertices in the graph. This is interesting since the time to check for the existence of a Hamiltonian path in the graph does not depend on the number of edges in the graph.

In [30] an interesting way to solve the 3CNF-SAT and HPP problem is presented using E-mails. The idea behind the solution is to represent each node by a person writing emails and each person knowing the email address of others. Then by parallely modifying the messages, we can find the solutions to the above problems in linear time. However, the above idea is theoretical and not practical.

In [29], AHNEPs for the recognition of context free languages are constructed. However the author could not provide bound for the size of AHNEPs used in recognition of context free languages which depends on the input alphabet only as it is reported for the generating hybrid network of evolutionary processors (HNEP). The following theorems are taken from [29].

Theorem 17 ([29]). *If L is a context free language and Γ is a non-deterministic push-down automaton with empty stack, without λ-transitions, $\Gamma = (Q, V, \Sigma, q_0, Z_0, \delta)$, then there exists an AHNEP Π such that $size(\Pi) = 3|Q| + 2|\sum| + 6$ and $L(\Pi) = L$.*

Theorem 18 ([29]). *If L is a context free language and Γ is a non-deterministic push-down automaton with empty stack, without λ-transitions, $\Gamma = (Q, V, \Sigma, q_0, Z_0, \delta)$, then there exists an AHNEP Π such that $size(\Pi) = |V| + 2|\sum| + 6$ and $L(\Pi) = L$.*

Theorem 19 ([29]). *Given a regular language L, let M be the minimum state deterministic finite state automaton that recognizes L and n be its number of states. Then there exists an AHNEP* Π *such that* $L = L(\Pi)$ *and* $size(\Pi) = 2n + 3$.

Theorem 20 ([29]). *Given a regular language L, let V be the set of symbols of L. Then there exists an AHNEP* Π *such that* $L = L(\Pi)$ *and* $size(\Pi) = |V| + 4$.

In [31], a variant of AHNEP called accepting network of splicing processors (ANSP in short) is defined. The model is same as AHNEP, except that here we use splicing rules, which is a biologically motivated operation [24] instead of simple point mutations. Other than this the rest of parameters for each node is same as in AHNEPs. Similar to AHNEPs, complexity classes based on ANSPs are defined and it is shown to solve HPP and 3-CNF-SAT problem in *linear* time using this model.

5 Network of Evolutionary Processors in Image Generation and Natural Language Processing

Here we show how to use NEP as a theoretical model for generating images. We also show how to use NEPs for natural language processing.

5.1 Pictural Network of Evolutionary Processors

Here we recall the definition of pictural network of evolutionary processors from [34].

Let V be an alphabet. A *picture* (or two-dimensional string) over the alphabet V is a two-dimensional array of elements from V. We denote the set of all pictures over the alphabet V by V^{**}, while the empty picture is denoted by ε. A two-dimensional language over V is a subset of V^{**}.

Let p be a picture in V^{**}. The number of rows and number of columns of p are denoted by $r(p)$ and $c(p)$ respectively. The pair $(r(p), c(p))$ is called the *size* of the picture p. The set of all pictures over V having the size (m, n) where $m, n \geq 1$ is denoted by $V^{m \times n}$. The symbol placed at the intersection of the ith row with j^{th} column of the picture p is denoted by $p(i, j)$ or equivalently $p_{i,j}$

A pictural network of evolutionary processors (PNEP) is similar to NEP except the following differences:

(1) Each node in a NEP has initially a multiset of strings (which may be empty), while each node in a PNEP has initially a multiset of pictures (which may be empty), i.e., each node in a PNEP has a multiset of two-dimensional strings before the start of the computation.

(2) The evolution rules in PNEP are of the following forms:

- $a \rightarrow b$, $a, b \in V$ (substitution rules),
- $a \rightarrow \varepsilon(r)$, $a \in V$ (row deletion rules),
- $a \rightarrow \varepsilon(c)$, $a \in V$ (column deletion rules),
- $\varepsilon \rightarrow a(r)(X)$, $a \in V$ (row insertion rules),
- $\varepsilon \rightarrow a(c)(Y)$, $a \in V$ (column insertion rules)

where $X \in \{u, d, -\}$ and $Y \in \{l, r, |\}$.

Instead of regular filters, the input and output filters of each node in a PNEP are defined by random context conditions as in HNEPs.

Here also the computation is an alternating sequence of evolutionary and communication step as in NEP, except that instead of one-dimensional strings we have two-dimensional strings.

5.1.1 Comparison with Other Picture Generating Models

We set:

LOC is the class of local picture languages [21].

REC is the class of recognizable picture languages [21].

RML is the class of regular matrix languages [38].

(R:R)KAL is the class of regular array languages considered in [39] with applications to generation of "Kolam" patterns [40].

BPL is the class of basic puzzle languages considered in [35].

$\mathcal{L}$(PNEP) is the class of two-dimensional languages generated by PNEPs.

The following theorem is taken from [34] which gives the comparison of PNEPs with other picture generating models.

Theorem 21 ([34]). *(i) The families $\mathcal{L}$(PNEP) and LOC are incomparable but not disjoint.*

(ii) The families $\mathcal{L}$(PNEP) intersects the families REC, RML and (R:R)KAL.

(iii) $\mathcal{L}$(PNEP) contains non-regular matrix languages.

(iv) $\mathcal{L}$(PNEP) intersects the picture class of RAG and BPG when we restrict to rectangular picture language only.

In [17], the concept of PNEP is extended to generate 2-dimensional and 3-dimensional pictures. The model is compared with other models of picture generation.

5.2 Analyzing Sentences using Network of Evolutionary Processors

In [18] the authors have used NEPs for introducing a formalization for implementing the recognition of correct sentences of natural language. This is done by specializing each node of a NEP for the labeling of different phrases of the sentence. It is shown that the system that can recognize strings can also analyze sentences of the natural language. The authors have shown how to construct a NEP which is able to decide whether or not a string or natural language is correct and analyze its structure. This is done using only lineal inputs and outputs, i.e., the sentences are analyzed by recognizing and labeling the internal pattern structure of the sentences rather than generating the syntax trees. The linguistic NEP for analyzing simple sentences recognize every word, make a map of its linguistic features, establish the beginning and the end to the complex units, gather the non-terminal units in phrases and finally give a lineal structural version of the whole sentence.

6 Network of Evolutionary Processors Based on Splicing Operations

The computational process in NEP is not exactly an evolutionary process in the Darwinian sense. But the rewriting operations might be interpreted as mutations and the filtering process might be viewed as a selection process. In the basic model of NEP, HNEP and AHNEP, the authors havenot considered the issue of recombination of DNA molecules. The central issue is to solve NP-complete problems efficiently in the biological way.

Since NEP is a bio inspired theoretical model of computation, it seems more appropriate to replace the point mutation operations in NEP by splicing [24], which is also a biologically motivated operation. This operation was introduced by Tom Head in [24] as a mathematical model of DNA recombination which consists of cutting DNA sequences and then pasting the fragments again, under the influence of restriction enzymes and ligases. There are several variants of splicing, such as splicing with permitting context, where we check for the presence of certain nucleotides in the DNA molecules before applying the splicing rules on them, splicing with forbidding context, where the DNA molecules are allowed to undergo splicing if it does not contain certain nucleotides. There exists several other variants of splicing. A comprehensive detail of various types of splicing operations can be found in [36].

6.1 Splicing and its Variants

Splicing is a biologically motivated operation. This operation was introduced by Tom Head [24] as a mathematical model of DNA recombination which consists of cutting DNA sequences and then pasting the fragments again, under the influence of restriction enzymes and ligases. Consider an alphabet V and two special symbols, #, \$, not in V. A *splicing rule* (over V) is a string of the form

$$r = u_1\#u_2\$u_3\#u_4,$$

where $u_1, u_2, u_3, u_4 \in V^*$. For a set of splicing rules R, *radius* of R is defined as $rad(R) = \max\{|x|, x = u_i, 1 \leq i \leq 4, \text{ for some } u_1\#u_2\$u_3\#u_4 \in R\}$. For a splicing rule $r = u_1\#u_2\$u_3\#u_4$ and strings $x, y, z \in V^*$ we write

$$\begin{aligned}(x,y) \vdash_r z \quad \text{iff} \quad & x = x_1u_1u_2x_2,\\ & y = y_1u_3u_4y_2,\\ & z = x_1u_1u_4y_2,\\ & \text{for some } x_1, x_2, y_1, y_2 \in V^*.\end{aligned}$$

The above operation is called single splicing, where we consider only the first string generated after applying the splicing rule. We say that we *splice* x, y at the *sites* u_1u_2, u_3, u_4, respectively and the result is z. In order to keep this operation as close as possible to reality, we consider the operation of the form

$$\begin{aligned}(x,y) \models_r (z,w) \quad \text{iff} \quad & x = x_1u_1u_2x_2,\\ & y = y_1u_3u_4y_2,\\ & z = x_1u_1u_4y_2,\\ & w = y_1u_3u_2x_2,\\ & \text{for some } x_1, x_2, y_1, y_2 \in V^*.\end{aligned}$$

where $r = u_1\#u_2\$u_3\#u_4$ is a splicing rule. This operation is called double splicing operation.

6.1.1 Splicing Rules with Permitting Context

Sometimes, DNA molecules undergo splicing operation under the influence of restriction enzymes, only if certain nucleotides are present in the DNA molecules. This scenario can be modeled in formal languages by using permitting context in each splicing rule, where we check for the presence of certain symbols in the strings, before applying the splicing rules on them. This variant of splicing rule is called splicing rule with permitting context [36].

Formally, a splicing rule with permitting context (over V) is a triple of the form $p = (r; D_1, D_2)$ with $r = u_1\#u_2\$u_3\#u_4$ being a splicing rule over V and D_1, D_2 being finite subsets of V^*. Note that here only the systems with finite components are considered. For x, y, z, $w \in V^*$ and $p = (r; D_1, D_2)$, we define:

$$(x, y) \models_p (z, w) \text{ iff } (x, y) \models_r (z, w),$$

with each element of D_1 appearing as a sub-string in x and each element of D_2 appearing as a sub-string in y; when $D_1 = \emptyset$ or $D_2 = \emptyset$, then no condition on x, respectively y, is imposed.

6.1.2 Splicing Rules with Forbidding Context

Similar to permitting context, we can use forbidding context in each splicing rule, where we check for the absence of certain symbols in the strings, before applying the splicing rules on them. Biologically, this represents the scenario where the DNA molecules undergo splicing operation, only if certain nucleotides are not present in the molecule [36].

A splicing rule with forbidding context (over V) is a triple of the form $p = (r; D_1, D_2)$ with $r = u_1\#u_2\$u_3\#u_4$ being a splicing rule over V and D_1, D_2 being finite subsets of V^*. Note that here only the systems with finite components are considered. For $x, y, z, w \in V^*$ and $p = (r; D_1, D_2)$, the following is defined:

$$(x, y) \models_p (z, w) \text{ iff } (x, y) \models_r (z, w),$$

with no element of D_1 appearing as a sub-string in x and no element of D_2 appearing as a sub-string in y; when $D_1 = \emptyset$ or $D_2 = \emptyset$, then no condition on x, respectively y, is imposed.

6.1.3 Simple Splicing Rules

Consider an alphabet V and a set $V' \subseteq V$. One can consider four types of languages over V^*, corresponding to the splicing rules of the forms

$$a\#\$a\#, \#a\$\#a, a\#\$\#a, \#a\$a\#,$$

where a is an arbitrary element of V'. The elements of V' are called markers. These four rules are called simple splicing rules of type $< 1, 3 >$, $< 2, 4 >$, $< 1, 4 >$ and $< 2, 3 >$ respectively [36].

Clearly splicing rules of types $< 1, 3 >$ and $< 2, 4 >$ yield the same result and for x, y, $z \in V^*$ and $a \in V'$ we obtain

$$(x, y) \vdash^a_{<(1,3)>} z \text{ iff } x = x_1 a x_2, y = y_1 a y_2, z = x_1 a y_2 \text{ , for some } x_1, x_2, y_1, y_2 \in V^*$$

For other types, the splicing is performed as follows:

$$(x, y) \vdash^a_{<(1,4)>} z \text{ iff } x = x_1 a x_2, y = y_1 a y_2, z = x_1 a a y_2 \text{ , for some } x_1, x_2, y_1, y_2 \in V^*$$

$$(x, y) \vdash^a_{<(2,3)>} z \text{ iff } x = x_1 a x_2, y = y_1 a y_2, z = x_1 y_2 \text{ , for some } x_1, x_2, y_1, y_2 \in V^*$$

6.2 Network of Evolutionary Processors with Splicing Rules

A network of evolutionary processors with splicing rules (NEPS in short) is a variant of NEP introduced in [11]. Here we use splicing rules instead of rewriting rules. In [11] it is shown that NEPS are computationally complete. It is also shown how to solve NP-complete problems in *linear* time using NEPS. The following theorems are taken from [11], which shows the power of NEPS as a language generating device as well as a computational device.

Theorem 22 ([11]). *Each recursively enumerable language can be generated by a complete NEPS of size 2 where the splicing rules are of type regular and filters are of type infinite regular languages.*

Theorem 23 ([11]). *Let $G = (\{1, 2, \ldots, n\}, E)$ be an arbitrary graph. We can find whether G is 3-colorable or not by using a NEPS of size $n + 1$ in $O(n)$ time, where the nodes of the NEPS are connected linearly.*

Theorem 24 ([11]). *Given a finite set X with $|X| = 3q$ and a collection C of 3-element subsets of X, we can find whether C contains an exact cover for X by using a linearly connected NEPS of size $n + q + 3$ in $O(n + q)$ time.*

In [11] linear time algorithms for subgraph isomorphism and Hamiltonian path problem are also given. All these problems are NP-complete problems and solving them in linear time by NEPS, shows the power of NEPS.

6.3 Network of Evolutionary Processors with Splicing Rules and Permitting/Forbidding Context

A network of evolutionary processors with splicing rules and permitting context (NEPSP in short) is a variant of HNEP, except that here we use splicing rules with permitting context instead of point mutations. In [10], it is shown that any recursively enumerable language can be generated by a NEPSP consisting only 2 nodes. It is also shown how to solve some well known NP-complete problems in *linear* time using this variant of HNEP. The following theorems are taken from [10].

Theorem 25 ([10]). *Each recursively enumerable language can be generated by a complete NEPSP of size two.*

Theorem 26 ([10]). *Let $(S = \{a_1, a_2, \ldots, a_n\}, F = \{F_1, F_2, \ldots, F_m\})$, be an (n, m)-instance of the CAP. It can be solved by a complete NEPSP of size $m + 2n + 2$ in $O(m + n)$ time.*

The Common Algorithmic Problem (CAP) is discussed in [25]: let S be a finite set and F be a non-empty family of subsets of S. Find the cardinality of a maximal subset of S which does not include any set belonging to F. The sets in F are called *forbidden sets*. We say that (F, S) is an $(card(S), card(F))$-instance of CAP. The problem is NP-complete [20]. Solving this problem in *linear* time shows the power of NEPSP as a computational device.

Similar to NEPSP, we have network of evolutionary processors with splicing rules and forbidding context (NEPSF in short) [12] where we use splicing rules with forbidding context instead of splicing rules with permitting context. Similar to NEPSP, in [12] it is shown that any recursively enumerable language can be generated by a NEPSF consisting only 2 nodes. This shows that this model is computationally complete. Also *linear* time algorithms to some well known NP-complete problems are proposed running on NEPSF.

6.4 Hybrid Network of Evolutionary Processors with Simple Splicing Rules

A hybrid network of evolutionary processors with simple splicing rules (HNEPSS in short) [9] is a variant of NEPSP. The parameters of each node is same as NEPSP, except that here we use simple splicing rules. Also each node has different mode of applying the rules, namely (l, l), (l, r), $(l, *)$, (r, l), (r, r), $(r, *)$, $(*, l)$, $(*, r)$ and $(*, *)$. Each node can have one of the modes of applying the rule out of the above nine modes. Since we are using simple splicing rules where we perform the splicing operation around (before or after depending upon the type) the same symbol a in both the strings involved in splicing, the action mode specifies the position of the symbol a in the strings on which splicing rule is applied. For example mode (l, r) implies that whenever we are using any splicing rule involving the symbol $a \in V'$, then a should be the leftmost symbol of the first string involved in the splicing and a should be the rightmost symbol of the second string involved in splicing. $\alpha = (r, *)$ implies that whenever we are using a splicing rule involving a symbol $a \in V'$, then a should occur at the rightmost position in the first string and a can be present at any position in the second string. Other values are interpreted accordingly. The definition of configuration, computation, evolutionary step and communication step is same as that in NEPSP.

In [9] it is shown that HNEPSS are computationally complete. The following result is taken from [9].

Theorem 27 ([9]). *Any recursively enumerable language over an alphabet T can be generated by a HNEPSS of size $27 + 3.card(T)$.*

The above theorem shows that the same HNEPSS over the alphabet T can be used to generate any recursively enumerable language over the alphabet T. Further in [9] it is stated that any recursively enumerable language can be generated by a HNEPSS of size 43. This is obtained by using a nice encoding scheme introduced in [41]. The idea behind the construction is as follows: We first define a HNEPSS consisting of 10 processors that converts the image $c(w) \in \{0, 1\}^*$ of a word $w \in V^*$ under a block coding c into w. Finally, the HNEPSS generating a language $L \subseteq V^*$ is found as follows. We first take the HNEPSS with 33 processors generating the language $c(L)$ and then join the HNEPSS that converts $c(w)$ to w.

The above result is interesting because by using simple splicing rules we cannot go beyond context free languages [36]. The power we get in HNEPSS is due to the parallel processing of strings, different modes of applying the rules, input filters and output filters.

6.5 Accepting Hybrid Network of Evolutionary Processors with Splicing Rules and Permitting/Forbidding Context

An accepting hybrid network of evolutionary processors with splicing rules and permitting context (AHNEPSP in short) [6] is an accepting device, used for solving membership decision problems. The basic model of AHNEPSP is same as that of NEPSP. Each node has some multiset of strings, multiset of axioms, finite set of splicing rules with permitting context, an input filter and an output filter defined by different types of random context conditions. One of the nodes is designated as the input node and one of the nodes is designated as output node.

Initially only the input node contains the string, for which we want to find the membership decision problem. All other nodes contain no strings. The computation proceeds as in NEPSP as an alternating sequence of evolutionary and communication step. A computation halts if one of the following two conditions holds:

(i) There exists a configuration in which the set of words existing in the output node is non-empty. In this case, the computation is said to be an *accepting computation*

(ii) There exist two consecutive identical configurations.

In the aforementioned cases the computation is said to be finite. The language accepted by a AHNEPSP Γ is

$$L(\Gamma) = \{w \in V^* |\text{ the computation of } \Gamma \text{ on } w \text{ is an accepting one }\},$$

where V is the alphabet of the network. Thus, AHNEPSP can be viewed as the automata model of NEPSP. The following results are taken from [6].

Theorem 28 ([6]). *Given a graph G with n vertices and m edges, we can find whether G is 3-colorable by using a complete AHNEPSP of size* $6n + 2$ *in* $O(m + n)$ *time.*

Theorem 29 ([6]). *Given 2 graphs* G_1 *and* G_2*, we can find whether* G_2 *is a subgraph of* G_1 *by using a complete AHNEPSP of size* $st + 2$ *in* $O(t + n)$ *time where s is the number of vertices in* G_1*, t is the number of vertices in* G_2 *and n is the number of edges in* G_2*.*

The above two theorems shows the capability of AHNEPSP as an accepting device.

Similarly, we have another accepting device called accepting network of evolutionary processors with splicing rules and forbidding context (AHNEPSF in short) [5]. In [5] it is shown how to solve Hamiltonian path problem and 3-CNF SAT problem in linear time using AHNEPSF.

6.6 Network of Evolutionary Processors with Parallel String Rewriting Rules

We say that a rule $x \to y$, where x, y are finite length strings over V^* is a *substitution rule* if none of x and y are ε; it is a *deletion rule* if $x \neq \varepsilon$ and $y = \varepsilon$. The set of all substitution and deletion rules over strings from an alphabet V are denoted by Sub_{V^*} and Del_{V^*}. Given a set of rules σ consisting of substitution or deletion rules as above and a word $w \in V^*$, we define the following *actions* of σ on w:

- If $\sigma \equiv \{x_i \to y_i, 1 \leq i \leq k\} \subseteq Sub_{V^*}$, then $\sigma^*(w) = \{u_0y_1u_1y_2 \ldots u_{k-1}y_ku_k;\ \exists u_j \in V^*, 0 \leq j \leq k$ where $w = u_0x_1u_1x_2 \ldots u_{k-1}x_ku_k$ and no substring of u_j is occurring at the left-hand side of any rule in $Sub_{V^*}\}$.
- If $\sigma \equiv \{x_i \to \varepsilon, 1 \leq i \leq k\} \subseteq Del_{V^*}$, then $\sigma^*(w) = \{u_0u_1 \ldots u_{k-1}u_k;\ \exists u_j \in V^*, 0 \leq j \leq k$ where $w = u_0x_1u_1x_2 \ldots u_{k-1}x_ku_k$ and no substring of u_j is a left-hand side of a rule in $Del_{V^*}\}$.

Here, we apply the rules parallely i.e., suppose in some node Y, we have a string $abcdefgh$, where a, b, c, d, e, f, g, h are symbols from some alphabet. Suppose Y has the substitution rules $abc \to ijk$ and $efg \to lmn$, where i, j, k, l, m, n are symbols over the same alphabet, then both the rules will be applied parallely in a single step, thus generating the string *ijkdlmnh*. A problem may come if overlapping takes place, i.e., suppose that instead of above 2 rules, we have the rules $abc \rightarrow ijk$ and $bcd \rightarrow lmn$. In this case, the left hand side of the 2 productions are overlapping, so we cannot apply both the rules parallely. Since, we will be having multiple copies of each string, we will be getting 2 strings, *ijkdefgh* and *almnefgh*. Similarly, deletion rules are also applied parallely.

A network of evolutionary processors with parallel string rewriting rules (NEPPS in short) is similar to NEPSP, except the following differences:

- Here we use string rewriting rules, which are applied parallely on the strings
- Here we use only one type of filter, namely the type (1) filter.

In [7] it is shown that using NEPPS, we can break DES (Data Encryption Standard) [37], which is the most widely used cryptosystem. It is also shown that the number of steps taken by the algorithm is less than the number of steps for breaking the same cryptosystem by other models.

Theorem 30 ([7]). *Given an arbitrary <plain-text, cipher-text> pair, where the cipher-text is obtained by encrypting the plain text using DES key, one can recover the DES key, using a linearly connected NEPPS consisting of 134 nodes in 268 steps.*

Similarly, in [8] it is shown that we can break a logic based cryptosystem, which was introduced in [27] and shown to be optimal in the sense that any cryptanalytic attack against it can be used to break other crypto systems as well.

7 Conclusion

The concept of network of evolutionary processors is based on mechanisms inspired from cell biology. A variation of these networks called hybrid network of evolutionary processors is also studied. These networks consists of nodes which are very simple processors and are able to perform just one type of point mutation (insertion, deletion or substitution of a symbol). These nodes are endowed with a filter which are defined by regular sets (for NEPs) and random context conditions (for HNEPs which seem to be close to the possibilities of biological implementation). In addition to this an accepting model (AHNEP) for solving decision problems are also defined. Also NEP are used as a theoretical model of picture generation and also for analyzing the structure of simple sentences in natural languages.

In all the 3 models, namely NEP, HNEP and AHNEP, recombination of DNA molecules is not considered. So we have introduced different variants of splicing operations, which is a biologically motivated operation in the above 3 models. It is shown that all these variants are computationally complete. Also they are efficient computational and accepting devices and can solve NP-complete problems in *linear* time. The usage of parallel string rewriting rules is also considered in NEP and it is found that the resulting model can break some well known cryptosystem in very less number of steps.

In [33] the authors have raised the following question: The authors are conscious that their mechanisms have likely no biological relevance. Then why to study them? The scientific community believe that by combining their knowledge about behavior of cell populations with advanced formal theories from computer science, they could try to define computational models based on the interacting molecular entities. To this aim we need to accomplish the followings:

(1) Understanding which features of the behavior of molecular entities forming a biological system can be used for designing computer networks with an underlying structure inspired from that of the biological system.

(2) Understanding how to control the data navigating in the networks via precise protocols.

(3) Understanding how to effectively design the networks.

The results mentioned in this paper suggest that these mechanisms might be a reasonable example of global computing due to the real and massively parallelism involved in molecular interactions. Therefore, they deserve a deep theoretical investigation as well as an investigation of biological limits of implementation.

We end this section by giving some directions for future research.

1. In [6] and [5] we have introduced accepting devices based on splicing rules with permitting/forbidding context. We have shown how to solve some well known NP-complete problems in *linear* time using these models. However we have not shown that this model is computationally complete. For this we have to show how to simulate the moves of a Turing machine on these models. This remains an open problem. Similarly, complexity classes based on these variants should be defined and compared with the classical complexity classes as done for AHNEP.

2. In [9] it is shown that we can generate any recursively enumerable language using a HNEPSS having 43 nodes. Can this number be reduced or can we put a bound on the number of nodes? This remains an open problem. Another open problem in this respect is to find whether there exists any recursively enumerable language which can be generated by a HNEPSS having less than 43 nodes.

3. In NEPS, NEPSP, NEPSF, HNEPSS, AHNEPSP and AHNEPSF we have considered different variants of splicing operations. However there is another variant of splicing operation which is called circular splicing [36]. It would be interesting to see the behavior of circular splicing on the above models.

4. In [34], PNEPs are just compared with other models of picture generation. It would be interesting to find whether PNEPs are computationally complete. Also the application of splicing rules instead of rewriting rules should be investigated in PNEPs.

5. The application of NEPs is investigated in [18]. However only simple sentences are considered. Effort should be made for carrying out more complex tasks like parsing of different sentence structures as well as complex syntactic pieces. This will require more complex architectures and the improvement of the parallelism of the system.

References

[1] http://psystems.disco.unimib.it.

[2] J. Castellanos, P. Leupold, and V. Mitrana. On the size complexity of hybrid networks of evolutionary processors. *Theoretical Computer Science*, 330(2):205–220, 2005.

[3] J. Castellanos, C. Martin-Vide, V. Mitrana, and J. Sempere. Networks of evolutionary processors. *Acta Informatica*, 39:517–529, 2003.

[4] J. Castellanos, C. Martin-Vide, V. Mitrana, and J. M. Sempere. Solving NP-complete problems with networks of evolutionary processors. In *Proc. IWANN 2001, LNCS 2084*, pages 621–628. Springer-Verlag, Berlin, 2001.

[5] A. Choudhary. Accepting hybrid network of evolutionary processors with splicing rules and forbidding context. accepted in IICAI 2005.

[6] A. Choudhary and K. Krithivasan. Accepting hybrid network of evolutionary processors with splicing rules and permitting context. submitted to ADCOM 2005.

[7] A. Choudhary and K. Krithivasan. Breaking DES using network of evolutionary processors with string rewriting rules. Communicated to International Journal of Computer Mathematics.

[8] A. Choudhary and K. Krithivasan. Breaking a logic based cryptosystem using network of evolutionary processors with parallel string rewriting rules. Accepted in CIT 2005.

[9] A. Choudhary and K. Krithivasan. Hybrid network of evolutionary processors with simple splicing rules. Communicated to International Journal of Applied Mathematics and Computing.

[10] A. Choudhary and K. Krithivasan. Network of evolutionary processors with splicing rules and permitting context. accepted in IPCAT 2005.

[11] A. Choudhary and K. Krithivasan. Network of evolutionary processors with splicing rules. In *In Proc. of IWINAC 2005, LNCS 3561*, pages 290–299, 2005.

[12] A. Choudhary and K. Krithivasan. Network of evolutionary processors with splicing rules and forbidding context. In *In Proc. of IWINAC 2005, LNCS 3561*, pages 300–309, 2005.

[13] E. Csuhaj-Varju, J. Dassow, J. Kelemen, and Gh. Păun. *Grammar Systems: A Grammatical Approach to Distribution and Cooperation*. Gordon and Breach, 1993.

[14] E. Csuhaj-Varju, C. Martin-Vide, and V. Mitrana. Hybrid networks of evolutionary processors are computationally complete. *Acta Informatica*, 41(4-5):257–272, 2005.

[15] E. Csuhaj-Varju and V. Mitrana. Evolutionary systems: A language generating device inspired by evolving communities of cells. *Acta Informatica*, 36:913–926, 2000.

[16] E. Csuhaj-Varju and A. Salomaa. Networks of parallel language processors. In *In Gh. Păun and A. Salomaa eds., New Trends in Formal Languages, LNCS 1218*, pages 299–318. Springer-Verlag, 1997.

[17] K. S. Dersanambika, K. G. Subramanian, and A. R. S. Mary. 2D and 3D pictural networks of evolutionary processors. In *Proc. of IWINAC 2005, LNCS 3562*, pages 92–101, 2005.

[18] G. B. Enguix and M. D. Jimenez-Lopez. Analysing sentences with networks of evolutionary processors. In *Proc. of IWINAC 2005, LNCS 3562*, pages 102–111, 2005.

[19] L. Errico and C. Jesshope. Towards a new architecture for symbolic processing. In *Artificial Intelligence and Information Control Systems of Robots '94 (I. Plander ed.)*, pages 31–40. World Scientific Publication, Singapore, 1994.

[20] M. Garey and D. Johnson. *Computers and Intractability: A Guide to the Theory of NP-Completeness*. Freeman, San Francisco, CA, 1979.

[21] D. Giammaressi and A. Restivo. Recognizable picture languages. *International Journal of Pattern Recognition and Artificial Intelligence*, 6:241–256, 1992.

[22] J. Hartmanis, P. M. Lewis II, and R. E. Stearns. Hierarchies of memory limited computations. In *Proc. 6th Annual IEEE Symp. on Switching Circuit Theory and Logical Design*, pages 179–190, 1965.

[23] J. Hartmanis and R. E. Stearns. On the computational complexity of algorithms. *Trans. Amer. Math. Soc*, 117:533–546, 1965.

[24] T. Head. Formal language theory and DNA: An analysis of the generative capacity of specific recombinant behaviors. *Bulletin of Mathematical Biology*, 49(6):737–759, 1987.

[25] T. Head, M. Yamamura, and S. Gal. Aqueous computing: Writing on molecules. In *Proc. of the Congress on Evolutionary Computation, 1999, IEEE Service Center, Piscataway, NJ*, pages 1006–1010, 1999.

[26] W. D. Hillis. *The Connection Machine*. MIT Press, Cambridge, 1985.

[27] J. Kari. A cryptosystem based on propositional logic. In J. Dassow and J. Kelemen, editors, *Machines, Languages and Complexity, 5th International Meeting of Young Computer Scientists*, LNCS 381, pages 210–219. Springer Verlag, 1988.

[28] S. Lloyd. Ultimate physical limits to computation. *Nature*, 406:1047–1054, 2001.

[29] F. Manea. Using ahneps in the recognition of context free languages. In *Proc. of the Workshop on Symbolic Networks, ECAI 2004*, 2005.

[30] F. Manea, C. Martin-Vide, and V. Mitrana. Solving 3CNF SAT and HPP in linear time using WWW. In *Proc. of MCU 2004*, pages 269–280, 2004.

[31] F. Manea, C. Martin-Vide, and V. Mitrana. Accepting networks of splicing processors. In *Proc. of CiE 2005, LNCS 3526*, pages 300–309, 2005.

[32] M. Margenstern, V. Mitrana, and M. Perez-Jimenez. Accepting hybrid networks of evolutionary processors. In *C. Ferreti, G. Mauri and C. Zandron eds. Proc. of DNA 10, LNCS 3384*, pages 235–246, 2004.

[33] C. Martin-Vide, V. Mitrana, M. J. Perez-Jimenez, and F. S. Caparrini. Hybrid networks of evolutionary processors. In *Proc. of GECCO 2003, LNCS 2723*, pages 401–412, 2003.

[34] V. Mitrana, K. G. Subramanian, and M. Tataram. Pictural networks of evolutionary processors. *Romanian Journal of Information Science and Technology*, 6(1-2):189–198, 2003.

[35] M. Nivat, A. Saoudi, K. G. Subramanian, R. Siromoney, and V. R. Dare. Puzzle grammars and context free array grammars. *International Journal of Pattern Recognition and Artificial Intelligence*, 5:663–676, 1991.

[36] Gh. Păun, G. Rozenberg, and A. Salomaa. *DNA Computing*. Springer-Verlag, 1998.

[37] A. Salomaa. *Public Key Cryptography*. Springer Verlag, 1990.

[38] G. Siromoney, R. Siromoney, and K. Krithivasan. Abstract families of matrices and picture languages. *Computer Graphics and Image Processing*, 1:284–307, 1972.

[39] G. Siromoney, R. Siromoney, and K. Krithivasan. Picture languages with array rewriting rules. *Information and Control*, 22:447–470, 1973.

[40] G. Siromoney, R. Siromoney, and K. Krithivasan. Array grammars and kolam. *Computer Graphics and Image Processing*, 3:63–82, 1974.

[41] R. Stiebe. Size considerations for HNEPs. manuscript.

Ashish Choudhary
Department of Computer Science and Engineering
Indian Institute of Technology, Madras
Chennai, India 600036
email: ashishc@cse.iitm.ernet.in

Formal Language Aspects of Natural Computing– 2006, pp. 143–153

Tissue-Like P Systems Communicating by Request*

Erzsébet Csuhaj-Varjú, Gheorghe Păun and György Vaszil

Abstract

We study tissue-like P systems using string-objects where the communication between the membranes is indicated by the occurrence of so-called query symbols in the string. We prove that these systems are computationally complete devices and they are computationally efficient in the sense that NP-complete problems can be solved in this framework in polynomial time.

1 Introduction

Membrane systems, or P systems, are distributed and parallel computing devices inspired by the functioning of the living cell [5]. The main ingredient of a P system is a hierarchically embedded structure of membranes. Each membrane encloses a region that contains objects and might also contain other membranes. There are rules associated to the regions describing the evolution and the movement of the objects which together correspond to a computation. For details on membrane systems, see the monograph [6] and consult the web-page `http://psystems.disco.unimib.it`.

An important direction of research, started in [4], is to study tissue-like P systems which, instead of an individual cell, correspond to groups of cells, like tissues or organs, interacting with each other either directly or with the use of the environment, but in any case, having the common property that the membrane structures are not necessarily described by a tree as the ones corresponding to individual cells.

In this paper we study tissue-like P systems where each membrane encloses a multiset of string objects. The evolution rules of the objects are represented by context-free rewriting rules which also describe the communication between the membranes by the help of special communication symbols, called query symbols, one such symbol corresponding to each region of the system.

When one or more query symbols are introduced in a string, then the rewriting of that string stops and the queries are satisfied by replacing the query symbols with strings which do not contain further query symbols from the region indicated by the query symbol, in all possible combinations. If no query symbol free string exists in the queried region, then the string containing the query disappears.

This model allows a biological interpretation if the strings are considered as descriptions of simple organisms, the query symbols as their "weak points", possibly infected or attacked by another organism, thus, the communication resembles some features of an infection or parasitism. Inspired by these resemblances, we call a communication of type i (infection) if after communicating the copies of the strings, the strings themselves remain in the region, while the communication is called of type p (parasitism), if after communication the communicated string itself disappears from its original region. The model is called an MPC system in short.

*Research supported by the Hungarian Scientific Research Fund, "OTKA", grants no. T042529 and F037567. The work of Gh. Păun was partially done during his stay as a visiting scientist at the Computer and Automation Research Institute, Hungarian Academy of Sciences, a Centre of Excellence in Information Technology, Computer Science, and Control, supported by project ICA1-CT-2000-70025, HUN-TING, WP5.

We note that MPC systems can also be considered as modified variants of parallel communicating (PC) grammar systems defined over multisets of strings. PC grammar systems are networks of grammars organized in a communicating system to generate a single language. The reader interested in the theory of grammar systems is referred to [1, 2].

In this paper we prove that MPC systems in both communication modes are able to determine non-context-free languages even with one membrane, while systems with seven membranes and working in the i mode are sufficient to describe all recursively enumerable languages. Then we show how MPC systems working in the p communication mode simulate MPC systems working in the i communication mode, thus we prove the computational completeness of systems working in communication mode p as well. Finally, we demonstrate the computational efficiency of the model by proving that the satisfiability of propositional formulas can be decided in this framework in polynomial time with respect to the number of variables and of clauses.

Study of problems motivated by the resemblance of these constructs to the above mentioned biological phenomena could be an interesting and promising research direction.

2 Prerequisites and Definitions

We introduce here only some notation and terminology. See [6] for more information on the elements of membrane computing.

The free monoid generated by an alphabet V with respect to the operation of concatenation is denoted by V^*, ε denotes the empty string. The families of context-free, and recursively enumerable languages are denoted by CF and RE, respectively. A multiset M over V^* is a mapping of V^* into the set $\mathbb{N}$ of non-negative integers. $M(w)$ is called the multiplicity of $w \in V^*$. The support of multiset M is the set $supp(M) = \{w \in V^* \mid M(w) > 0\}$. We say that a word $w \in V^*$ is an element of M, denoted by $w \in M$, if $w \in supp(M)$. M is called finite, if its support is finite. We say that M is empty, denoted as in the case of sets by $\emptyset$, if its support is empty, $supp(M) = \emptyset$. The union of two multisets over V^* is defined as $M' = (M_1 \cup M_2)$ over V^* where for all $w \in V^*$, $M'(w) = M_1(w) + M_2(w)$. We will use the notation of double brackets $M = \{\{x \mid x$ satisfies condition $C\}\}$ for describing (or enumerating) the elements of a multiset M.

Definition 1. A tissue-like P system with string objects and context-free rules communicating by request, an MPC system in short, is a construct

$$\Pi = (N, K, T, (M_1, R_1), \ldots, (M_m, R_m), i_o),\ m \geq 1,$$

where N, K, T are pairwise disjoint alphabets with $K = \{Q_1, \ldots, Q_m\}$ (the elements of K are called *query symbols* and they are associated with the m components of Π, we denote $V = (N \cup K \cup T)$; $M_1, \ldots, M_m$ are finite multisets over $(N \cup T)^*$; $R_1, \ldots, R_m$ are finite sets of context-free rules of the form $A \to u$, with $A \in N$ and $u \in V^*$; and finally, $i_o \in \{1, 2, \ldots, m\}$ is the *output* region of Π. Note that in contrast to the customary notation, the communication channels between the regions are given implicitly by the rewriting rules.

Instead of presenting the full, complicated formal details of the definition of the work of MPC systems, we describe it as follows. The computation starts in the initial configuration $(M_1, \ldots, M_m)$ and passes from a given configuration $(M'_1, \ldots, M'_m)$, consisting of multisets of strings over $N \cup T$ placed in the m regions of the system, to another configuration $(M_1'', \ldots, M_m'')$ in the following way. Each string from each multiset M_i' which can be rewritten according to the rules from R_i, $1 \leq i \leq m$, has to be rewritten. This means the use of one applicable rule from R_i, non-deterministically chosen, for each string (at the level of each string, the rewriting is sequential). The strings which cannot be rewritten (no rule can be applied to them) remain unchanged. If no query symbol is introduced in any of the strings, then the resulting multisets are M_i'', $1 \leq i \leq m$. Note that the rewriting of strings is maximally parallel in the sense that all strings which can be rewritten must be rewritten, and that the process is non-deterministic, i.e. the choice of rules and the places where the rules are applied can lead to several possible new multisets of strings.

If, as the result of the rewriting process, a query symbol appears in any of the strings, then a communication is performed: Each symbol Q_j present in a string from region i is replaced with all strings from region j which do not contain query symbols. If region j contains several strings without query symbols, then each of them is used, hence the string from region i is replicated (with the occurrence of Q_j replaced with strings from region j). If there are several query symbols in the same string from region i, then all of them are replaced (we also say that they are *satisfied*) at the same time, in all possible combinations. If a query symbol Q_j cannot be satisfied (region j contains no string without query symbols), then the string containing Q_j is removed. In this way, all query symbols introduced by the rewriting rules disappear: they are either satisfied or they disappear together with the string which contain them (in the case when they cannot be satisfied). The multisets obtained after the communication are $M_1'', \ldots, M_m''$, constituting the next configuration of the system.

We call an MPC system *i*-communicating if copies of the requested strings are communicated to the requesting components, and *p*-communicating if after replacing the query symbols with the requested strings, these strings are removed from the multiset associated to the queried region.

Let us denote the transitions from one configuration to another, $(M_1', \ldots, M_m')$ to $(M_1'', \ldots, M_m'')$, by $(M_1', \ldots, M_m') \Rightarrow_X (M_1'', \ldots, M_m'')$ with $X = i$ and $X = p$ for *i*-communicating and *p*-communicating systems, respectively. The language generated by the system consists of all terminal strings produced in region i_o during any possible computation in Π.

$$L_X(\Pi) = \{x \in T^* \mid (M_1, \ldots, M_m) \Rightarrow_X^* (M_1', \ldots, M_m') \text{ and } x \in M_{i_o}'\}$$

for $X \in \{i, p\}$, where $\Rightarrow_X^*$ denote the reflexive and transitive closure of $\Rightarrow_X$.

The families of all languages described by MPC systems of degree at most $m \geq 1$, with context-free rewriting rules and i mode or p mode of communication, is denoted by $iMPC_mCF$ or $pMPC_mCF$, respectively. If we use systems of an arbitrary degree, then we replace the subscript m with $*$.

3 The Power of MPC Systems

The following relations are direct consequences of the definitions.

$$CF \subseteq XMPC_nCF \subseteq XMPC_{n+1}CF \subseteq XMPC_*CF \subseteq RE$$

for $X \in \{i, p\}$, and all $n \geq 1$. Surprisingly enough, the leftmost inclusion is proper in both communication modes even for $n = 1$.

Lemma 1. *$XMPC_1CF - CF \neq \emptyset$ for $X \in \{i, p\}$.*

Proof: Let us consider first $\Pi = (\{S, A\}, \{Q_1\}, \{a, b\}, (M_1, R_1), 1)$, a p-communicating MPC system (of degree 1) with

$$M_1 = \{\{S, A, b\}\}, \text{ and}$$
$$R_1 = \{S \to aSa,\ S \to Q_1,\ A \to A,\ A \to Q_1Q_1\}.$$

We start with three strings in the system, S, A, and b. From S we can generate a^nSa^n, for any $n \geq 0$, then we replace S with Q_1. If at the same time we have produced Q_1Q_1 from A, then we simultaneously get a^nba^n and bb, both of them included in $L(\Pi)$, and the computation stops.

If A is still present, then we get the strings a^nAa^n and a^nba^n, and A and b are "consumed". The string a^nAa^n will eventually lead to $a^nQ_1Q_1a^n$, and hence to $a^na^nba^na^nba^na^n$, which is in $L(\Pi)$.

If from A we produce Q_1Q_1 while both a^nSa^n and b are present, then we obtain four strings: $a^nSa^na^nSa^n$, a^nSa^nb, ba^nSa^n, bb (and b is no longer present). The first three strings will be rewritten by the rule $S \to aSa$ for a number of steps, and eventually the rule $S \to Q_1$ is used. The query symbol will be either replaced by a string which contains again S, hence the process continues, or by bb. Consequently, if any terminal string is obtained on this path, then it contains the substring bb.

Thus, $L_p(\Pi) \cap a^+ba^+ba^+ = \{a^{2n}ba^{2n}ba^{2n} \mid n \geq 1\}$, which is not a context-free language.

For the i-communicating case, let $\Pi' = (\{S_1, S_2\}, \{Q_1\}, \{a, b, c\}, (M_1', R_1'), 1)$ where

$$M_1' = \{\{S_1, bS_2, a\}\},$$
$$R_1' = \{S_1 \to Q_1Q_1Q_1S_1, S_1 \to Q_1Q_1Q_1, S_2 \to Q_1S_2, S_2 \to Q_1c\}.$$

We do not give a detailed explanation, but the reader might easily check that the set of terminal strings which can be produced by Π' is not context-free since $L_i(\Pi') \cap a^+ba^+cba^+cba^+c = \{a^{3n+3j}ba^{n+1}cba^{n+1}cba^{n+1}c \mid n, j \geq 0\}$ which is not a context-free language. □

MPC systems with context-free rules are computationally universal; they characterize the class of recursively enumerable languages in both communication modes, even with a limited number of components.

Before proving this result, we recall the notion of a two-counter machine from [3]. A *two-counter machine* $TCM = (T \cup \{Z, B\}, E, R)$ is a 3-tape Turing machine where T is an *alphabet*, E is a set of *internal states* with two distinct elements $q_0, q_F \in E$, and R is a set of *transition rules*. The machine has a read-only input tape and two semi-infinite storage tapes (the counters). The alphabet of the storage tapes contains only two symbols, Z and B (blank), while the alphabet of the input tape is $T \cup \{B\}$. R contains transition rules of the form $(q, x, c_1, c_2) \to (q', e_1, e_2)$ where $x \in T \cup \{\varepsilon\}$ corresponds to the symbol scanned on the input tape in state $q \in E$, and $c_1, c_2 \in \{Z, *\}$ correspond to the symbols scanned on the storage tapes. If $c_i = Z$, then the symbol scanned on the ith counter tape is Z, if $c_i = *$, then the symbol scanned is either Z or B. By a rule of this form, M enters state $q' \in E$, and the counters are modified according to $e_1, e_2 \in \{-1, 0, +1\}$. If $x \in T$, then the machine was scanning x on the input tape, and the head moves one cell to the right; if $x = \varepsilon$, then the machine performs the transition irrespective of the scanned input symbol, and the reading head does not move.

The symbol Z appears initially on the cells scanned by the storage tape heads and may never appear on any other cell. An integer t can be stored by moving a tape head t cells to the right of Z. A stored number can be incremented or decremented by moving the tape head right or left. The machine is capable of checking whether a stored value is *zero* or not by looking at the symbol scanned by the storage tape heads. If the scanned symbol is Z, then the value stored in the corresponding counter is *zero*. Note that although we do not allow to explicitly check the non-emptiness of the counters which is allowed in [3], this feature can be simulated: After successfully decrementing and incrementing a counter, the stored value is not altered, but the machine can be sure that the scanned symbol is B. A word $w \in T^*$ is accepted by the two counter machine if the input head has read the last non-blank symbol on the input tape, and the machine is in the accepting state q_F. Two-counter machines are computationally complete; they are just as powerful as Turing-machines, see [3].

Lemma 2. $iMPC_7CF = RE$.

Proof: We only prove $RE \subseteq iMPC_7CF$, the reverse inclusion follows from the Church thesis. To this aim, let us consider a recursively enumerable language $L \subseteq T^*$ and a two-counter machine $TCM = (T \cup \{Z, B\}, E, R)$, in the above form, accepting the language L. We construct an i-communicating MPC system generating L. Let

$$\Pi = (N, K, T, C_{sel}, C_{gen}, C_{c_1}, C_{c_2}, C_{ind}, C_{ch}, C_{del}, sel),$$

where (M_α, R_α) is denoted as C_α for $\alpha \in \{sel, gen, c_1, c_2, ind, ch, del\}$.

Let $D = \{[q, x, c_1, c_2, q', e_1, e_2] \mid (q, x, c_1, c_2) \to (q', e_1, e_2) \in R\}$, and let us define for any $\alpha = [q, x, c_1, c_2, q', e_1, e_2] \in D$, the following notations: $State(\alpha) = q$, $Read(\alpha) = x$, $NextState(\alpha) = q'$, and $Store(\alpha, i) = c_i$, $Action(\alpha, i) = e_i$ for $i = 1, 2$.

The general idea of the simulation is to represent the states and the transitions of TCM with nonterminals of D and the values of the counters by strings of nonterminals containing as many A symbols as the value stored in the given counter.

Let

$$N = \{S, [I]_7, [F], E, A, T, T', T'', T''', T^{iv}\} \cup \{S_i, E_i, T_i \mid 1 \le i \le 6\} \cup \{\alpha_i \mid \alpha \in D, 1 \le i \le 7\},$$

and let the rules be defined as follows.

$$\begin{aligned} M_{sel} &= \{\{[I]_7\}\}, \\ R_{sel} &= \{[I]_7 \to \alpha_1 \mid \alpha \in D, State(\alpha) = q_0\} \cup \{E \to \varepsilon, T'' \to \varepsilon\} \cup \\ &\quad \{\alpha_7 \to \beta_1 \mid \alpha, \beta \in D, NextState(\alpha) = State(\beta)\} \cup \\ &\quad \{\alpha_7 \to [F] \mid \alpha \in D, NextState(\alpha) = q_F\} \cup \\ &\quad \{\alpha_i \to \alpha_{i+1} \mid \alpha \in D, 1 \le i \le 6\} \cup \{[F] \to [F], [F] \to Q_{ch}Q_{del}\}. \end{aligned}$$

This region keeps track of the current state of the simulated two-counter machine and also selects the transition to be simulated. The symbol $[I]_7$ is used to initialize the system by introducing one of the initial transition symbols of the form $[q_0, x, c_1, c_2, q', e_1, e_2]_1$ where q_0 is the initial state. It also produces the result of the computation when after simulating the entering of the counter machine into the final state (that is, after the appearance of the nonterminal $[F]$), it receives the strings produced in the regions corresponding to C_{ch},

C_{del}, and erases the occurrences of the nonterminals T'', E which, if the simulation was successful, produces a terminal word accepted by the two-counter machine.

$$M_{gen} = \{\{S\}\},$$
$$R_{gen} = \{S \rightarrow Q_{sel}\} \cup \{\alpha_i \rightarrow \alpha_{i+1} \mid \alpha \in D, 1 \leq i \leq 5\} \cup \{[F] \rightarrow \varepsilon\} \cup$$
$$\{\alpha_6 \rightarrow xQ_{c_1}Q_{c_2}S \mid Read(\alpha) = x\}.$$

This region generates the string accepted by the counter machine by adding the symbol $Read(\alpha)$ for each $\alpha \in D$ chosen in the selector region. It also appends, during the simulation of each step of the two-counter machine, the strings denoting the counter contents, so at the end of the computation the system will be able to make sure that no "mistake" was made during the simulation.

For $i = 1, 2$, let

$$M_{c_i} = \{\{T\}\},$$
$$R_{c_i} = \{T \rightarrow T', T' \rightarrow T'', T'' \rightarrow Q_{sel}, A \rightarrow Q_{ind}, T''' \rightarrow Q_{ind}, T^{iv} \rightarrow T,$$
$$E \rightarrow E_1\} \cup \{\alpha_3 \rightarrow \delta_\alpha T''' \mid \alpha \in D, \delta_\alpha = AA \text{ if } Action(\alpha, i) = +1,$$
$$\delta_\alpha = A \text{ if } Action(\alpha, i) = 0, \delta_\alpha = \varepsilon \text{ if } Action(\alpha, i) = -1\}.$$

These regions maintain strings representing the contents of the two counters. After the selection of a transition symbol in the region corresponding to C_{sel}, they execute the action required by the chosen transition symbol by adding AA, A, or ε to the counter string and then deleting one A and T''' by rewriting it to Q_{ind}. The simulation can only be successful if exactly one A and the symbol T''' is rewritten. This is ensured by region C_{ind}. If there is a string obtained after the two queries which contain only a number of A or E symbols and one T^{iv} symbol, then the simulation of the actions required by the chosen transition was successful. If a counter is empty, this construction also forbids the successful execution of the decrement instruction since this would introduce E_1 in the counter strings.

The rules of the region supporting the work of the counters C_{c_1} and C_{c_2} is defined as follows.

$$M_{ind} = \{\{E_1, T_1\}\},$$
$$R_{ind} = \{E_i \rightarrow E_{i+1} \mid 1 \leq i \leq 4\} \cup \{E_5 \rightarrow E, E \rightarrow E_6, E_6 \rightarrow E_1\} \cup$$
$$\{T_i \rightarrow T_{i+1} \mid 1 \leq i \leq 5\} \cup \{T_6 \rightarrow T^{iv}, T^{iv} \rightarrow T_1\}.$$

Finally, let

$$M_{ch} = \{\{S_1\}\},$$
$$R_{ch} = \{S_1 \rightarrow Q_{sel}, S_6 \rightarrow S_1, S_i \rightarrow S_{i+1} \mid 2 \leq i \leq 5\} \cup \{[F] \rightarrow \varepsilon\} \cup$$
$$\{\alpha_1 \rightarrow \delta_1\delta_2S_2 \mid \alpha \in D, \delta_j = Q_{c_j} \text{ if } Store(\alpha, j) = Z,$$
$$\text{or } \delta_j = \varepsilon \text{ otherwise}\}, \text{ and let}$$
$$M_{del} = \{\{S_1\}\},$$
$$R_{del} = \{S_1 \rightarrow Q_{sel}, S_6 \rightarrow S_1, S_i \rightarrow S_{i+1} \mid 2 \leq i \leq 5\} \cup$$
$$\{\alpha_1 \rightarrow S_2 \mid \alpha \in D\} \cup \{[F] \rightarrow Q_{gen}, A \rightarrow \varepsilon, T \rightarrow \varepsilon\}.$$

These regions help to check the correctness of the simulation. C_{ch} collects the counter contents whenever the simulated transition requires an empty counter. If the counter is really empty, it receives a string of the form $zT''v$ where $z, v \in \{E\}^*$. If there are any A symbols present, then C_{sel} will not be able to remove them at the end of the computation, thus, no terminal strings will be produced. Region C_{del} waits for the appearance of the symbol $[F]$ signaling the end of the simulation, after which it receives the sentential forms of C_{gen}, and then deletes all occurrences of the nonterminals T, A. If there is an obtained string which besides the nonterminal E contains only terminal letters, then C_{sel} will be able to produce a terminal word, one accepted by the simulated two-counter machine by concatenating the sentential forms obtained from C_{ch} and C_{del}, and then erasing all occurrences of the letters T'' and E.

Table 1: Components of Π in the proof of Lemma 2 with $\alpha = [q, x, Z, B, q', +1, -1]$.

	C_{sel}	C_{gen}	C_{c_1}	C_{c_2}	C_{ind}	C_{ch}	C_{del}
0	β_7	wS	T	$A...T$	E_1, T_1	zS_1	S_1
1	α_1	wQ_{sel}	T'	$A...T'$	E_2, T_2	zQ_{sel}	Q_{sel}
1C	α_1	$w\alpha_1$	T'	$A...T'$	E_2, T_2	$z\alpha_1$	α_1
2	α_2	$w\alpha_2$	T''	$A...T''$	E_3, T_3	$zQ_{c_1}S_2$	S_2
2C	α_2	$w\alpha_2$	T''	$A...T''$	E_3, T_3	$zT''S_2$	S_2
3	α_3	$w\alpha_3$	Q_{sel}	$A...Q_{sel}$	E_4, T_4	$zT''S_3$	S_3
3C	α_3	$w\alpha_3$	α_3	$A...\alpha_3$	E_4, T_4	$zT''S_3$	S_3
4	α_4	$w\alpha_4$	AAT'''	$A...T'''$	E_5, T_5	$zT''S_4$	S_4
5	α_5	$w\alpha_5$	$AQ_{ind}T'''$	$Q_{ind}...T'''$	E, T_6	$T''S_5$	S_5
5C	α_5	$w\alpha_5$	AET''', AT_6T'''	$E...T''', T_6...T'''$	E, T_6	$T''S_5$	S_5
6	α_6	$w\alpha_6$	AEQ_{ind} AT_6Q_{ind}	$E...Q_{ind}$ $T_6...Q_{ind}$	E_6, T^{iv}	$zT''S_6$	S_6
6C	α_6	$w\alpha_6$	AEE_6, AT_6E_6 $\underline{AET^{iv}}, AT_6T^{iv}$	$E...E_6, T_6...E_6$ $\underline{E...T^{iv}}, T_6...T^{iv}$	E_6, T^{iv}	$zT''S_6$	S_6
7	α_7	$wxQ_{c_1}Q_{c_2}S$	AET	$E...T$	E_1, T_1	$zT''S_1$	S_1
7C	α_7	$wxAETE...TS$	AET	$E...T$	E_1, T_1	$zT''S_1$	S_1

Instead of giving a detailed proof of the correctness of our construction, we demonstrate the work of the MPC system in Table 1 by indicating a possible transition sequence of Π while simulating an instruction of the two-counter machine TCM. The instruction is represented by $\alpha_1 = [q, x, Z, *, q', +1, -1]_1$ chosen in region C_{sel} in the first step. This indicates that the first counter should be empty which requirement is satisfied since region C_{c_1} contains a string containing zero A symbols. The nonterminals α_1 and α_3 are communicated to the other membranes where the required actions on the counters are executed. The emptiness of the first counter is made sure by C_{ch} when in the second step the counter region C_{c_1} is queried. Note that we only indicate those strings (at regions $C_{gen}, C_{c_1}, C_{c_2}, C_{ch}$) which contribute to the successful simulation of the two-counter machine, and furthermore, T and $A...T$ represent strings over $\{A, E, T\}^*$ containing zero, or more than one A symbols, respectively. □

Now we show that p-communicating systems can simulate the work of i-communicating ones.

Lemma 3. *$iMPC_nCF \subseteq pMPC_{n+3}CF$, for any $n \geq 1$.*

Proof: Let $\Pi = (N, K, T, (M_1, R_1), \ldots, (M_n, R_n), 1)$ be an MPC system of degree n. We construct $\Pi' = (N', K', T, (M'_1, R'_1), \ldots, (M'_{n+3}, R'_{n+3}), 1)$ of degree $n + 3$, such that $L_p(\Pi') = L_i(\Pi)$.

Let $N' = N \cup \{S_1, S_2, S_3, S_4, S_5, S_6\}$, and let the rules of Π' be defined as

$$M'_i = M_i \cup \{\{S_1, S_2\}\},$$
$$R'_i = R_i \cup \{S_1 \to Q_i, S_3 \to Q_i, S_2 \to Q_{n+1}, S_4 \to Q_{n+1}\}$$

for all $1 \leq i \leq n$, and

$$M'_{n+1} = \{\{S_1, S_2, S_3, S_4\}\}, \quad R'_{n+1} = \{S_3 \to Q_{n+2}, S_4 \to Q_{n+3}\},$$
$$M'_{n+2} = \{\{S_3, S_4, S_5, S_6\}\}, \quad R'_{n+2} = \{S_5 \to Q_{n+1}, S_6 \to Q_{n+3}\},$$
$$M'_{n+3} = \{\{S_1, S_2, S_5, S_6\}\}, \quad R'_{n+3} = \{S_1 \to Q_{n+1}, S_2 \to Q_{n+2}\}.$$

The additional membranes of Π' work as "suppliers" of symbols, when region $n + 1$ is queried, it returns either S_1, S_2 or S_3, S_4. These symbols are used in the other regions to keep a copy of the contents of each multiset after any communication simulating this way the communication behaviour of Π. □

We synthesize the above results in the following hierarchy theorem.

Theorem 1. *$CF \subset XMPC_1CF \subseteq iMPC_7CF = pMPC_{10}CF = XMPC_*CF = RE$ for $X \in \{i, p\}$.*

4 The Efficiency of MPC Systems

We show now how the possibility of MPC systems to create exponentially many strings in a linear time can be used for solving computationally hard problems in a polynomial time. The framework is customary in membrane computing (see [6] and the formal approaches from [7] in particular). Briefly speaking, we work with *confluent* systems, constructed in a *semi-uniform* manner. This means that starting from a given instance of a decision problem, we construct in a polynomial time (by a Turing machine) an MPC system which always stops in a known number of steps, in spite of a possible non-deterministic behavior, and all computations give the same result, which is the answer to the problem (that is, the system is also *sound* and *complete*). The result will be obtained in the output component; for decision problems, the answer will be *yes* if and only if in a specified step this component will contain a specified string.

We illustrate this strategy by solving the satisfiability problems for propositional formulas.

Theorem 2. *The satisfiability of a propositional formula in conjunctive normal form consisting of m clauses and n variables can be decided by a p-communicating MPC system in $2mn + 2m + 3n + 1$ steps, $n, m \geq 1$.*

Proof: Let us consider a propositional formula $\gamma = C_1 \wedge \cdots \wedge C_m$, in conjunctive normal form, consisting of m clauses and n variables, $C_j = y_{j,1} \vee \cdots \vee y_{j,k_j}$, $1 \leq j \leq m$, where $y_{j,i} \in \{x_l, \neg x_l \mid 1 \leq l \leq n\}$, $1 \leq i \leq k_j$.

For each i, $1 \leq i \leq n$, let us denote by $true(\gamma, t_i)$ the string $C_{j_1} C_{j_2} \ldots C_{j_s}$ indicating the clauses in γ which contain x_i, and by $true(\gamma, f_i)$ the string $C_{j_1} C_{j_2} \ldots C_{j_s}$ of clauses in γ which contain $\neg x_i$ (therefore, these strings are over the alphabet $\{C_i \mid 1 \leq i \leq m\}$).

We construct the MPC system (of degree $3n + 2m + 5$, with the components labeled with $0, 1.1, 1.2, 1.3, \ldots, i.1, i.2, i.3, \ldots, n.1, n.2, n.3, 1, 1', \ldots, m, m', m+1, p.1, p.2, p.3$)

$$\Pi = (N, K, T, (M_0, R_0), (M_{1.1}, R_{1.1}), \ldots, (M_{p.3}, R_{p.3}), m+1)$$

where $N = \{A_i \mid 1 \leq i \leq mn + 3n + 2m + 1\} \cup \{A'_i,\ A''_i \mid 1 \leq i \leq 2n\} \cup \{T_i,\ F_i \mid 1 \leq i \leq n\} \cup \{C_i \mid 1 \leq i \leq m\} \cup \{B, D_1, D_2, D_3, D_4, D_5, D_6\}$, $T = \{d\}$,
and

$$\begin{aligned} M_0 &= \{\{B\}\}, & M_{i.1} &= M_{i.2} = M_j = M_{j'} = M_{m+1} = \{\{A_1\}\}, \\ M_{i.3} &= \{\{A'_1,\ A''_1\}\}, & M_{p.1} &= \{\{D_1, D_2, D_3, D_4\}\}, \\ M_{p.2} &= \{\{D_3, D_4, D_5, D_6\}\}, & M_{p.3} &= \{\{D_1, D_2, D_5, D_6\}\}, \end{aligned}$$

for all $1 \leq i \leq n$, $1 \leq j \leq m$. The rules are defined as follows. Let $R_0 = \emptyset$ and

$$\begin{aligned} R_{1.1} &= \{A_1 \rightarrow Q_0,\ B \rightarrow T_1 B\}, \\ R_{1.2} &= \{A_1 \rightarrow Q_0,\ B \rightarrow F_1 B\}, \\ R_{1.3} &= \{A'_1 \rightarrow A'_2,\ A''_1 \rightarrow A''_2,\ A'_2 \rightarrow Q_{1.1},\ A''_2 \rightarrow Q_{1.2}\}. \end{aligned}$$

For all i, $2 \leq i \leq n-1$,

$$\begin{aligned} R_{i.1} &= \{A_j \rightarrow A_{j+1} \mid 1 \leq j \leq 2(i-1)\} \cup \{A_{2i-1} \rightarrow Q_{i-1.3},\ B \rightarrow T_i B\}, \\ R_{i.2} &= \{A_j \rightarrow A_{j+1} \mid 1 \leq j \leq 2(i-1)\} \cup \{A_{2i-1} \rightarrow Q_{i-1.3},\ B \rightarrow F_i B\}, \\ R_{i.3} &= \{A'_j \rightarrow A'_{j+1},\ A''_j \rightarrow A''_{j+1} \mid 1 \leq j \leq 2i-1\} \cup \\ &\quad \{A'_{2i} \rightarrow Q_{i.1},\ A''_{2i} \rightarrow Q_{i.2}\}, \text{ and} \\ R_{n.1} &= \{A_j \rightarrow A_{j+1} \mid 1 \leq j \leq 2(n-1)\} \cup \{A_{2n-1} \rightarrow Q_{n-1.3},\ B \rightarrow T_n\}, \\ R_{n.2} &= \{A_j \rightarrow A_{j+1} \mid 1 \leq j \leq 2(n-1)\} \cup \{A_{2n-1} \rightarrow Q_{n-1.3},\ B \rightarrow F_n\}, \\ R_{n.3} &= \{A'_j \rightarrow A'_{j+1},\ A''_j \rightarrow A''_{j+1} \mid 1 \leq j \leq 2n-1\} \cup \\ &\quad \{A'_{2n} \rightarrow Q_{n.1},\ A''_{2n} \rightarrow Q_{n.2}\} \cup \\ &\quad \{T_i \rightarrow true(\gamma, t_i),\ F_i \rightarrow true(\gamma, f_i) \mid 1 \leq i \leq n\}. \end{aligned}$$

Furthermore,

$$\begin{aligned} R_1 &= \{A_j \rightarrow A_{j+1} \mid 1 \leq j \leq 3n\} \cup \{A_{3n+1} \rightarrow Q_{n.3},\ C_1 \rightarrow Q_{p.1}\}, \\ R_{1'} &= \{A_j \rightarrow A_{j+1} \mid 1 \leq j \leq 3n+1\} \cup \{A_{3n+2} \rightarrow Q_1\}, \\ R_i &= \{A_j \rightarrow A_{j+1} \mid 1 \leq j \leq 3n + n(i-1) + 2(i-1)\} \cup \\ &\quad \{A_{3n+n(i-1)+2(i-1)+1} \rightarrow Q_{i-1},\ C_i \rightarrow Q_{p.1}\}, \\ R_{i'} &= \{A_j \rightarrow A_{j+1} \mid 1 \leq j \leq 3n + n(i-1) + 2(i-1) + 1\} \cup \\ &\quad \{A_{3n+n(i-1)+2(i-1)+2} \rightarrow Q_i\}, \text{ for all } i,\ 2 \leq i \leq m, \end{aligned}$$

$$R_{m+1} = \{A_j \to A_{j+1} \mid 1 \leq j \leq 3n + m(n+2)\} \cup \{A_{3n+m(n+2)+1} \to Q_m\} \cup \{D_i \to d \mid 1 \leq i \leq 6\}, \text{ and}$$

$$R_{p.1} = \{D_3 \to Q_{p.2}, D_4 \to Q_{p.3}\},\ R_{p.2} = \{D_5 \to Q_{p.1}, D_6 \to Q_{p.3}\},$$
$$R_{p.3} = \{D_1 \to Q_{p.1}, D_2 \to Q_{p.2}\}.$$

The system works as follows. All symbols A_i, A'_i, A''_i are counters, they are used for the synchronization of the computation. Membranes $i.1$, $i.2$, $i.3$ are used for expanding the variable x_i, so that in region $n.3$ we get all 2^n truth-assignments for the n variables in the form of strings $\alpha_1\alpha_2 \ldots \alpha_n$, with $\alpha_i \in \{T_i, F_i\}$, with T_i indicating the value *true* and F_i indicating the value *false*, $1 \leq i \leq n$. After obtaining all 2^n truth-assignments $\alpha_1\alpha_2 \ldots \alpha_n$, with $\alpha_i \in \{T_i, F_i\}$, $1 \leq i \leq n$, in region $n.3$ we use the rules $T_i \to true(\gamma, t_i)$, $F_i \to true(\gamma, f_i)$. Each string has length n, hence we will have n steps to perform in order to replace each T_i, F_i by the sequence of satisfied clauses. (Notice that it is possible that the same T_i, F_i satisfies none, one, or several clauses, hence the strings over $\{C_j \mid 1 \leq j \leq m\}$ we obtain can be of any length between zero – when a truth-assignment satisfies no clause – to nm – the case of a truth-assignment where each T_i, F_i satisfies all clauses.)

After these $3n$ steps, we pass to examining all the obtained strings, checking whether at least one exists which contains all C_i, $1 \leq i \leq m$. During the $3n$ steps, components $j, j', j = 1, 2, \ldots, m$, have just increased the counters. In step $3n + 1$, in region 1 we use the rule $A_{3n+1} \to Q_{n.3}$, which brings all strings from region $n.3$ into region 1. In all first $3n + 1$ steps, component $1'$ has increased the counters. In step $3n + 2$ we can use the rule $C_1 \to Q_{p.1}$ in region 1 (and this is obligatory for each string which contains the symbol C_1) and, simultaneously, the rule $A_{3n+2} \to Q_1$ in component $1'$. In this way, all strings which contain at least one occurrence of C_1 will contain now the query symbol $Q_{p.1}$, while the strings which do not contain C_1 remain unchanged (hence they contain no query symbol). Therefore, all strings which do not contain the query symbol are erased from region 1, and moved to region $1'$. Simultaneously, all strings which were "protected" by the symbol $Q_{p.1}$ will replace $Q_{p.1}$ by a nonterminal D_k for some $k = 1, 2, 3, 4$ and will remain in region 1 for n more steps guaranteeing the replacement of all possible occurrences of C_1.

Then membrane 2 will request all strings from region 1. The process is iterated, checking in the same manner whether $C_2, C_3, \ldots, C_m$ are present. At the end, we have in component m the strings which correspond to the truth-assignments which have satisfied all clauses – maybe none, if no such a truth-assignment exists. If any string survives the checking phase, then in step $3n + mn + 2m + 1$, when membrane $m + 1$ uses the rule $A_{3n+mn+2m+1} \to Q_m$, we can see in region $m + 1$ whether the formula is satisfiable: a string arrives here if and only if the formula is satisfiable. This string is rewritten to a terminal word in at most nm steps, thus the whole computation requires $2mn + 3n + 2m + 1$ steps.

Notice that the conditions we have stated at the beginning of this section in order to claim that we have solved the problem are fulfilled: the system can be constructed in a polynomial time, it is confluent, sound and complete. □

5 Closing Remarks

There are several interesting and important topics for further study. For example, the MPC system constructed in the previous proof depends only on n and m, with one exception, the rules $T_i \to true(\gamma, t_i)$, $F_i \to true(\gamma, f_i)$, for $1 \leq i \leq n$, from $R_{n.3}$, which directly depend on the instance of the problem we are handling. Therefore, if we consider these rules as an input to the system, then the construction can be considered uniform. If this convention is not allowed, we might still obtain a uniform solution if the problem instance is codified by means of strings which have to be introduced in the initial configuration of the system in a specified component (added to the multiset of that component). The construction of such a system remains as an open problem.

Interesting questions concern the size complexity properties of MPC systems, for example, to find the minimal number of components that i-communicating or p-communicating MPC systems need to compute any recursively enumerable language.

The comparison of MPC systems and other variants of tissue-like P systems, and the comparison of parallel communicating grammar systems and MPC systems are also of interest.

References

[1] E. Csuhaj-Varjú, J. Dassow, J. Kelemen, Gh. Păun: *Grammar Systems: A Grammatical Approach to Distribution and Cooperation.* Gordon and Breach, London, 1994.

[2] J. Dassow, Gh. Păun, G. Rozenberg: Grammar Systems. In: G. Rozenberg, A. Salomaa, eds., *Handbook of Formal Languages* **2**, Chapter 4, Springer-Verlag, Berlin, 1997.

[3] P.C. Fischer: Turing machines with restricted memory access. *Information and Control*, **9** (1966), 364–379.

[4] C. Martín-Vide, Gh. Păun, J. Pazos, A. Rodríguez-Patón: Tissue P systems. *Theoretical Computer Science* **296 (2)** (2003), 295–326.

[5] Gh. Păun: Computing with membranes. *Journal of Computer and System Sciences* **61 (1)** (2000), 108–143 (and Turku Center for Computer Science-TUCS Report 208, November 1998, `www.tucs.fi`).

[6] Gh. Păun: *Membrane Computing: An Introduction.* Springer-Verlag, Berlin, 2002.

[7] M. Pérez-Jiménez, A. Romero-Jiménez, F. Sancho-Caparrini: Complexity classes in cellular computation with membranes. *Natural Computing* **2 (3)** (2003), 265–285.

Erzsébet Csuhaj-Varjú and György Vaszil
Computer and Automation Research Institute
Hungarian Academy of Sciences
Kende utca 13–17, H-1111 Budapest, Hungary
email: {csuhaj,vaszil}@sztaki.hu

Gheorghe Păun
Institute of Mathematics of the Romanian Academy
PO Box 1-764, 014700 Bucureşti, Romania, and
Research Group on Natural Computing
Department of Computer Science and AI, University of Sevilla
Avda. Reina Mercedes s/n, 41012 Sevilla, Spain
email: george.paun@imar.ro, gpaun@us.es

Formal Language Aspects of Natural Computing– 2006, pp. 155–160

Array-Rewriting P systems with Conditional Communication

S. Hemalatha, K.S. Dersanambika, K.G. Subramanian and C. Sri Hari Nagore

Abstract

In [1], the notion of conditional communication has been introduced in (string) rewriting P systems. In this note this notion is extended to and incorporated in array-rewriting P systems [2].

1 Introduction

P systems, with string-objects and string-processing rules as known in formal language theory, have been considered [3] and investigated extensively. Among several variations of these systems, conditional communication is a feature introduced in [1]. On the other hand rewriting P systems with string-objects have been extended to array-rewriting P systems with array-objects and the power of such systems has been studied in [2]. Here a beginning is made to incorporate the feature of conditional communication in array-rewriting P systems of [2]. The resulting P system, called array-rewriting P system with conditional communication, enables us to generate all solid rectangles and all solid squares over a single letter.

2 Basic Definitions

We recall the notions of string rewriting P systems with conditional communication [1] and array-rewriting P systems [2].

An extended (string) rewriting P system (of degree $m \geq 1$)with conditional communication [1] is a construct

$$\Pi = (V, T, \mu, M_1, M_2, \ldots, M_m, R_1, P_1, F_1, R_2, P_2, F_2, \ldots, R_m, P_m, F_m)$$

where: V is the alphabet of the system, $T \subseteq V$ is the terminal alphabet, μ is a membrane structure with m membranes (injectively labeled by $1, 2, \ldots, m$), M_i are finite languages over V, representing the strings initially present in the regions i, $i = 1, 2, \ldots, m$ of the system, R_i are finite sets of context-free rules over V present in regions i, $i = 1, 2, \ldots, m$ of the system, P_i are permitting conditions and F_i are forbidding conditions associated with regions i, $i = 1, 2, \ldots, m$. The conditions can be of the following forms:

1. empty : No restriction is imposed on strings, they either freely exit the current membrane or enter any of the directly inner membranes ; An empty permitting condition is denoted by (true,α), $\alpha \in \{in, out\}$, and an empty forbidding condition by (false,notα), $\alpha \in \{in, out\}$.

2. symbols checking : each P_i is a set of pairs (a, α), $\alpha \in \{in, out\}$ for $a \in V$ and each F_i is a set of pairs $(b$, not$\alpha)$, $\alpha \in \{in, out\}$ for $b \in$ V: a string w can go to a lower membrane only if there is a pair $(a, in) \in P_i$ with $a \in alph(w)$, and for each

$(b, notin) \in F_i$ we have $b \notin alph(w)$; similarly for the string to go out of membrane i , it is necessary to have $a \in alph(w)$ for at least one pair $(a, out) \in P_i$ and $b \notin alph(w)$ for all $(b, notout) \in F_i$

3. substrings checking : each P_i is a set of pairs (u, α), $\alpha \in \{in, out\}$ for $u \in V^+$ and each F_i is a set of pairs $(v, not\alpha)$, $\alpha \in \{in, out\}$ for $v \in V^+$: a string w can go to a lower membrane only if there is a pair $(u, in) \in P_i$ with $u \in Sub(w)$, and for each $(v, notin) \in F_i$ we have $v \notin Sub(w)$; similarly for the string to go out of membrane i , it is necessary to have $u \in Sub(w)$ for at least one pair $(u, out) \in P_i$ and $v \notin Sub(w)$ for all $(v, notout) \in F_i$.

Thus we have conditions of the types empty, symb, sub_k, respectively, where k is the length of the longest string in all P_i, F_i; when no upper bound is imposed we replace the subscript by $*$. A system is said to be non-extended if $V = T$.

The transitions in a system as above are defined in the following way. In each region,each string which can be rewritten by a rule from that region is rewritten. The rule to be applied and the symbol rewritten by it are non-deterministically chosen. Each string obtained in this way is checked against the conditions P_i, F_i from that region. If it fulfills the requested conditions, then it will be immediately sent out of the membrane or to an inner membrane, if any exists; if it fulfills both in and out conditions, then it is sent either out of the membrane or to a lower membrane, non-deterministically choosing the direction- and non-deterministically choosing the inner membrane in the case when several directly inner membrane exist. If a string does not fulfill any condition, or it fulfills only in conditions and there is no inner membrane, then the string remains in the same region. If a string cannot be rewritten(this can be the case with strings from other membranes), then it is directly checked against the communication conditions, and as above, it leaves the membrane (or remains inside forever) depending on the result of this checking.

The language generated by the above system is denoted by $L(\Pi)$. The family of all languages $L(\Pi)$ generated by the system Π of degree at most $m \geq 1$ with permitting conditions of type α and forbidding conditions of type β, is denoted by $[E]LSP_m(rw, \alpha, \beta)$, α, $\beta \in \{empty, symb\} \bigcup \{sub_k | k \geq 2\}$. If the degree of the systems is not bounded, then the subscript m is replaced by $*$

Example 1. *[1] Consider the system*

$$\Pi = (\{a, b\}, \{a, b\}, [_1[_2]_2]_1, \{a\}, \emptyset, R_1, P_1, F_1, R_2, P_2, F_2)$$

$R_1 = \{a \to bb\}$, $P_1 = \{(true, in), (true, out)\}$ $F_1 = \{(a, notin), (a, notout)\}$ $R_2 = \{b \to a\}$, $P_2 = \{(true, out)\}$, $F_2 = \{(b, notout)\}$ *generates the language* $L(\Pi) = \{b^{2^n} | n \geq 0\}$

An array is a mapping $A : Z^2 \to V \bigcup \sharp$ with a finite support, supp(A), where $supp(A) = \{v \in Z^2 | A(v) \neq \sharp\}$. In order to specify an array it is sufficient to specify the set $(v, A(v))$, for $v \in supp(A)$.

For example the L-shaped angle with equal arms is given by the sets,

$$\{((0, 0), a), ((1, 0), a), ((2, 0), a), ((3, 0), a), ((0, 1), a), ((0, 2), a),))0, 3), a)\}$$

all the other elements have $\sharp$ symbol. The empty array is denoted by λ and the set of all arrays over V is given by $V^{*2} = V^{+2} \cup \{\lambda\}$. Any subset of V^{*2} is an array language.

An array production $\wp$ over V is a triple $\wp = (W, A, B)$, where W is a finite subset of Z^2 and A, B are arrays with the supports included in W. For two arrays C, D over V and a production $\wp$ as above, we write $C \Rightarrow_{\wp} D$ if D can be obtained by replacing a sub array of C identical to A with B; all pixels of W which are blank in A should be blank also in C.

A regular array production is one of the following forms:

$A\sharp \rightarrow aB$, $\sharp A \rightarrow Ba$, $\begin{array}{c}\sharp\\ A\end{array} \rightarrow \begin{array}{c}B\\ a\end{array}$, $\begin{array}{c}A\\ \sharp\end{array} \rightarrow \begin{array}{c}a\\ B\end{array}$, $A \rightarrow a$, where A, B are non-terminals and a is a terminal.

An Array P System [2] is a construct

$$\Pi = (V, T, \mu, F_1, \ldots, F_m, R_1, \ldots, R_m, i_0),$$

where: V is the alphabet of the system, $T \subseteq V$ is the terminal alphabet, μ is a membrane structure with m membranes (labeled injectively by 1,2,. . . ,m), F_i are finite set of arrays over V, representing the arrays initially present in the regions i, $i = 1, 2, \ldots, m$, of the system, $R_1, R_2, \ldots, R_m$ are finite sets of array-rewriting rules over V of the form $\mathbf{A} \rightarrow \mathbf{B}(tar)$; and i_0 is the output elementary membrane of μ. The system is said to be non-extended if $V = T$.

A computation in an array P system is defined in the same way as in a string rewriting P system with the successful computations being the halting ones. A computation is successful only if it stops, a configuration is reached where no rule can be applied to the existing arrays, The result of a halting computation consists of the arrays composed only of symbols from T placed in the membrane with label i_0 in the halting configuration.

Example 2. *Consider the non-extended context-free system [2]*

$$\Pi = \left(\{a\}, \{a\}, \sharp, [_1[_2[_3]_3]_2]_1, \begin{array}{c}a\\ a\end{array}, \phi, \phi, R_1, R_2, 3\right)$$

$$R_1 = \left\{ \begin{array}{cc} & \sharp\\ \sharp & a\end{array} \rightarrow \begin{array}{cc} & a\\ \sharp & a\end{array} (in)\right\},$$

$$R_2 = \left\{ \begin{array}{cc} a & \sharp\\ \sharp & \end{array} \rightarrow \begin{array}{cc} a & a\\ \sharp & \end{array} (out), \begin{array}{ccc} a & \sharp & \sharp\\ \sharp & & \end{array} \rightarrow \begin{array}{ccc} a & a & a\\ \sharp & & \end{array} (in)\right\},$$

$$R_3 = \phi$$

The above system generates all L-shaped angles with equal arms, each arm being of length at least three.

3 Array - Rewriting P systems with Conditional Communication

A beginning is made here to incorporate the notion of conditional communication in array-rewriting P systems. An extended array-rewriting P system (of degree $m \geq 1$) with conditional communication is a construct

$$\Pi = (V, T, \mu, M_1, \ldots, M_m, R_1, P_1, F_1, \ldots, R_m, P_m, F_m),$$

where V is the alphabet of the system, $T \subseteq V$ is the terminal alphabet, μ is a membrane structure with m membranes (injectively labeled by 1,2,. . . ,m),

$M_1, \ldots, M_m$ are finite set of arrays over V, representing the arrays initially present in the regions 1,2,...,m of the system, $R_1, R_2, \ldots, R_m$ are finite sets of array - rewriting rules over V present in regions 1,2,...,m of the system, P_i are permitting conditions and F_i are forbidding conditions associated with region i, $1 \leq i \leq m$.

The conditions can be in the forms empty, symbol checking, subarray checking, which are defined analogous to the corresponding forms in the string case. The difference is that the objects are arrays. Thus we have the conditions of the types empty, symb, subarr in all P_i, F_i. The transitions in an array-rewriting P system with conditional communication are analogous to the string case. But the result of a halting computation is as defined for array-rewriting P systems.

The set of all arrays computed by an array-rewriting P system Π with conditional communication is is denoted by $[E]AL(\Pi)$ with E being omitted when the system is non-extended. The family of array languages generated by systems as above is denoted by $[E]ALP_m(arw, \alpha, \beta), \alpha, \beta \in \{empty, symb, subarr\}$ We illustrate with an example.

Example 3. *Consider the non-extended system*

$$\Pi = \left(\{a\}, \{a\}, \sharp, [_1]_1, \begin{matrix} a & \\ a & a \end{matrix}, R_1, P_1, F_1, 1\right)$$

$R_1 = \left\{a\sharp \to aa, \begin{matrix} \sharp \\ a \end{matrix} \to \begin{matrix} a \\ a \end{matrix}\right\}$, $P_1 = \{(a, in)\}$, $F_1 = \{a, notin\}$, *generates all L-shaped pictures over a.*

Proposition 1. *An array-rewriting P system with conditional communication in the class $EALP_3(symb, subarr)$ generates all solid rectangles over a single symbol a.*

Consider the extended system

$$\Pi_1 = (\{a, b, c, d\}, \{a\}, \sharp, [_1[_2[_3]_3]_2]_1, db, \phi, \phi, R_1, P_1, F_1, R_2, P_2, F_2, R_3, P_3, F_3, 3)$$

$$R_1 = \{d\sharp \to dd, b\sharp \to db, b\sharp \to dc\}, \quad P_1 = \{(c, in)\},$$

$$F_1 = \left\{\left(\begin{matrix} d\,\sharp \\ d\,d \end{matrix}, notin\right), \left(\begin{matrix} d\,\sharp \\ a\,d \end{matrix}, notin\right), \left(\begin{matrix} \sharp\,d \\ d\,d \end{matrix}, notin\right), \left(\begin{matrix} \sharp\,d \\ d\,a \end{matrix}, notin\right), \right\},$$

$$R_2 = \left\{\begin{matrix} \sharp \\ d \end{matrix} \to \begin{matrix} d \\ a \end{matrix}, \begin{matrix} \sharp \\ c \end{matrix} \to \begin{matrix} c \\ d \end{matrix}, \begin{matrix} \sharp \\ c \end{matrix} \to \begin{matrix} b \\ d \end{matrix}, b \to a\right\}, \quad P_2 = \{(a, in), (b, out)\},$$

$$F_2 = \left\{\left(\begin{matrix} d & \sharp \\ d & d \end{matrix}, \alpha\right), \left(\begin{matrix} \sharp\,d \\ d\,d \end{matrix}, \alpha\right), \left(\begin{matrix} \sharp & d \\ d & a \end{matrix}, \alpha\right), \left(\begin{matrix} d & \sharp \\ a & d \end{matrix}, \alpha\right), \alpha \in \{notin, notout\}\right\},$$

$$R_3 = \{d \to a\}, \quad P_3 = \{a, in\}, \quad F_3 = \{(a, notin)\}$$

Proposition 2. *An array-rewriting P system with conditional communication in the class $EALP_3(symb, subarr)$ generates all solid squares over a single symbol a.*

Consider the extended system,

$$\Pi_1 = \left(\{a, b, c, d\}, \{a\}, \sharp, [_1[_2[_3]_3]_2]_1, \begin{matrix} d & b \\ a & d \end{matrix}, \phi, \phi, R_1, P_1, F_1, R_2, P_2, F_2, R_3, P_3, F_3, 3\right)$$

$$R_1 = \{d\sharp \rightarrow ad, b\sharp \rightarrow dc\}, \quad P_1 = \{(c, in)\},$$

$$F_1 = \left\{\left(\begin{matrix} d\,\sharp \\ d\,d \end{matrix}, notin\right), \left(\begin{matrix} d\,\sharp \\ a\,d \end{matrix}, notin\right), \left(\begin{matrix} \sharp\,d \\ d\,d \end{matrix}, notin\right), \left(\begin{matrix} \sharp\,d \\ d\,a \end{matrix}, notin\right), \right\},$$

$$R_2 = \left\{\begin{matrix} \sharp \\ d \end{matrix} \rightarrow \begin{matrix} d \\ a \end{matrix}, \begin{matrix} \sharp \\ c \end{matrix} \rightarrow \begin{matrix} c \\ d \end{matrix}, \begin{matrix} \sharp \\ c \end{matrix} \rightarrow \begin{matrix} b \\ d \end{matrix}, b \rightarrow a\right\}, \quad P_2 = \{(a, in), (b, out)\},$$

$$F_2 = \left\{\left(\begin{matrix} d\,\sharp \\ d\,d \end{matrix}, \alpha\right), \left(\begin{matrix} \sharp\,d \\ d\,d \end{matrix}, \alpha\right), \left(\begin{matrix} \sharp\,d \\ d\,a \end{matrix}, \alpha\right), \left(\begin{matrix} d\,\sharp \\ a\,d \end{matrix}, \alpha\right), \alpha \in \{notin, notout\}\right\},$$

$$R_3 = \{d \rightarrow a\}, \quad P_3 = \{a, in\}, \quad F_3 = \{(a, notin)\}$$

Corollary 1. *The class of array languages* $[E]ALP_3(symb, subarr)$ *intersects the class of regular array languages [4].*

We note [4] that rectangles and squares are generated by Regular array grammars.

Corollary 2. *The class of array languages* $ALP_3(symb, symb)$ *intersects the class of array rewriting P systems [2]*

This is a consequence of example 3, which can be generated by Array rewriting *P* systems.

Proposition 3. $FIN \subset ALP_1(symb, symb)$*, where FIN denotes the finite array languages L, consisting of only finite number of arrays.*

If L is a finite array language over an alphabet V, then $L = \{p_1, p_2, \ldots, p_k\}$ is generated by $AP_1(symb, symb)$ system. For, $\Pi = (V, V, M_1, P_1, F_1)$, where $M_1 = L = \{p_1, p_2, \ldots, p_k\}$, $P_1 = \{a, in\}$, $F_1 = (a, notin)$, for some $a \in p_i$, $1 \leq i \leq k$. Also the language L of example 3 in $ALP_1(symb, symb)$ is not finite.

Proposition 4. $ALP_2(symb, symb) - RAL \neq \phi$*, where RAL denotes the class of all regular array languages.*

Again this is a consequence of example 3.

4 Conclusion

It remains to examine other properties of the array-rewriting *P* systems with conditional communication.

References

[1] P. Bottoni, A. Labella, C. Martin-vide, Gh. Păun: Rewriting P Systems with Conditional Communication. *Lecture Notes in Computer Science, Springer, Berlin* **2300** 325-353 (2002).

[2] R. Ceterchi, M. Mutyam, Gh. Păun, K.G. Subramanian: Array - Rewriting P Systems. *Natural Computing* **2** 229-249, (2003).

[3] Gh. Păun: Membrane Computing, An Introduction. *Springer*, (2002).

[4] Y. Yamamoto, K. Morita, K. Sugata: Context-sensitivity of two-dimensional regular array grammars. In: P.S.P. Wang (ed) Array Grammars, Patterns and Recognizers **18** 17-41, (1989).

S. Hemalatha and Sri Hari Nagore
Department of Applied Sciences and Humanities
Madras Institute of Technology
Anna University,
Chennai 600 044, India.
email: hemrav66@rediffmail.com, cshari@annauniv.edu

K.S. Dersanambika
Department of Mathematics
Fatima Mata National College
Kollam, Kerala 629 004, India
email: dersanapdf@yahoo.com

Subramanian
Department of Mathematics,
Madras Christian College,
Chennai 600 059, India
email: kgsmani1948@yahoo.com

Formal Language Aspects of Natural Computing– 2006, pp. 161–169

Descriptional Complexity of h-alternating Finite Automata

J. Kavitha and G. Sethuraman

Abstract

State complexity is a descriptive complexity measure for regular languages. We investigate the state complexity of some operations on regular languages. In particular, we consider the concatenation, complementation, intersection, difference and shuffle operations of regular languages represented by an h-Alternating Finite Automata(h-AFA). Most of the shown bounds are tight in the exact number of states. That is the number is sufficient in the worst case.

1 Introduction

Regular languages and finite automata are one of the oldest topics in computer science. They have been extensively studied since the 1950s. Their most celebrated application has been lexical analysis in programming language compilation. Other applications include circuit design, text editing, pattern matching, parallel processing, image generation and compression, DNA computing etc. These applications and implementations of regular languages motivate the study of descriptional complexity of regular languages. Some aspects of this area are now intensively investigated [7, 8]. One such aspects is the state complexity of regular languages and their operations.

The state complexity of a regular languages is the minimal number of states of Automaton representing the language. The state complexity of an operation on regular languages is a function that associates the sizes of the automata representing the operands of the operation to the minimal number of states of the automata representing the resulting language. State complexity is a natural measurement of operations on regular languages. It also gives a lower bound for the time and space complexity of those operations.

Some early results concerning the state complexity of regular languages can be found in [3, 4]. The state complexity of some operations on regular languages was investigated in [9, 10]. Yu et al [4] were the first to systematically study the complexity of regular language operations. Their paper was followed by several article investigating the state complexity of finite languages operations and unary languages operations [5, 6]. The nondeterministic state complexity of regular languages operations was studied by Holzer and Kutrib in [10].

The notion of alternation is a natural generalization of nondeterminism. It received its first formal treatment by Chandra, Kozen and Stockmeyer in 1976 [1]. In this paper they proved that the AFA are precisely as powerful as DFA as far as language recognition is concerned. They also have shown that there exist k-state AFA such that any equivalent complete DFA has atleast 2^{2^k} states. A more detailed treatment of AFA and their operations can be found in [2].

An h-AFA is a modified version of an AFA. An h-AFA is a s-AFA of [4] with a generalized acceptance condition. In this paper, we investigate the state complexity of some operations on regular languages for an h-AFA.

2 h-Alternating Finite Automata

In an NFA, if there are two or more possible transitions for the current state and the current input symbol, the outcomes of all the possible computation for the remaining input word are logically OR-ed. The definition of AFA extends this idea and allows arbitrary boolean operations in place of the $''\bigvee''$ operation.

More specifically, an AFA works in the following way: When the automaton reads an input symbol $'a'$ in a given state q, it will activate all states of the automaton to work on the remaining part of the input in parallel. Once the states have completed their tasks, q will compute its value by applying a boolean function on those results and pass on the resulting value to the state by which it was activated. A word w is accepted if the starting state computes the value of 1. It is rejected otherwise.

Denote by the symbol B the two-element Boolean algebra $B = (\{0, 1\}, \bigvee, \bigwedge, \bar{}, 0, 1)$. Let Q be a set. Then B^Q is the set of all mappings of Q into B. Note that $u \in B^Q$ can be considered as a vector of $|Q|$ entries, indexed by elements of Q, with each entry being from B. For $u \in B^Q$ and $q \in Q$, We write u_q to denote the image of q under u.

An h-AFA is a quintuple (Q, Σ, g, h, f), where

Q is the finite set of states,

Σ is the input alphabet,

$g : Q \times \Sigma \times B^Q \rightarrow B$ is the transition function where B is the two element Boolean algebra, B^Q refers to all boolean functions from Q to B,

$h : B^Q \rightarrow B$ is the accepting Boolean function,

$F \subseteq Q$ is the set of final states.

$g_q : \Sigma \times B^Q \rightarrow B$ to denote that g is restricted to state q,

That is $g_q(a, u) = g(q, a, u)$ for $a \in \Sigma$, $u \in B^Q$, $q \in Q$.

We write g_Q to denote the function from $\Sigma \times B^Q$ to B^Q that is obtained by combining the functions g_q, $q \in Q$

$g_Q \equiv (g_q)_{q \in Q}$

Let $u \in B^Q$. We use u_q, $q \in Q$, to denote that u is restricted to q.

We extend the definition of g of an h-AFA to a function: $Q \times \Sigma^* \times B^Q \rightarrow B$ as follows:

$g(q, \lambda, u) = u_q$

$g(q, aw, u) = g(q, a, g(w, u))$

for all $q \in Q, a \in \Sigma, w \in \Sigma^*, u \in B^Q$

Given an h-AFA $A = (Q, \Sigma, h, F, g)$ for $w \in \Sigma^*$, w is accepted by A if and only if $h(g(w, f)) = 1$, where f is the characteristic vector of F, that is $f_q = 1$ iff $q \in F$.

Example 1. *We define an h-AFA* $A = (Q, \Sigma, h, F, g)$, *where*

$Q = \{q_1, q_2\}$,

$\Sigma = \{a, b\}$,

$F = \{q_2\}$,

$h(q_1, q_2) = q_1 \bigwedge \overline{q_2}$

and g *is given by*

state	a	b
q_1	$q_1 \bigvee \overline{q_2}$	$q_1 \bigwedge \overline{q_2}$
q_2	$\overline{q}_1 \bigwedge \overline{q_2}$	$\overline{q_1} \bigvee \overline{q}_2$

Let $w = aaab$. *Then* w *is accepted by* A *as follows :*

$h(g(aaab, f))$

$$
\begin{aligned}
&= h(g(a, g(aab, f)))\\
&= h(g(a, g(a, g(ab, f))))\\
&= h(g(a, g(a, g(a, g(b, f)))))\\
&= h(g(a, g(a, g(a, g(b, (0, 1))))))\\
&= h(g(a, g(a, g(a, (0, 1)))))\\
&= h(g(a, g(a, (0, 0))))\\
&= h(g(a, (1, 1)))\\
&= h((1, 0))\\
&= 1.
\end{aligned}
$$

3 Operations on regular Languages

We start our investigations with Boolean operations. In the case, when the finite automaton is deterministic it is well known that in the worst case the Boolean operations union, intersection and complementation have a state complexity of mn, mn and m respectively. However, the state complexity of an h-AFA operations is essentially much less. At first we consider the concatenation of two different languages defined by an h-AFA.

3.1 Concatenation Operation

Theorem 1. *For any Positive integers m, n let A be an m-state h-AFA and B be an n-state h-AFA. Then $m+n$ states are sufficient in the worst case for an h-AFA to accept the language $L(A)L(B)$.*

Proof: Let A be an h-AFA $A = (Q', \Sigma_1, g', h', F')$ accepts the language $L(A)$ with m states.

Let B be an h-AFA $B = (Q'', \Sigma_2, g'', h'', F'')$ accepts the language $L(B)$ with n states such that $Q' \bigcap Q'' = \phi$ and $\Sigma_1 \bigcap \Sigma_2 = \phi$.

We construct an $m+n$ state h-AFA $C = (Q, \Sigma, h, F, g)$ which accepts $L(A)L(B)$ as follows.

$$
\begin{aligned}
Q &= Q' \bigcup Q'',\\
\Sigma &= \Sigma_1 \bigcup \Sigma_2,\\
F &= F' \bigcup F'',\\
h &= h' \bigwedge h''
\end{aligned}
$$

and g is defined as

$$
g_q(a, u) = \begin{cases}
g'_q(a, u) & \text{if } q \in Q', a \in \Sigma_1\\
g''_q(a, u) & \text{if } q \in Q'', a \in \Sigma_2\\
0 & \text{if } q \notin F', a \in \Sigma_2\\
1 & \text{if } q \in F', a \in \Sigma_2\\
q & \text{if } q \in Q'', a \in \Sigma_1
\end{cases}
$$

Claim : $L(C) = L(A)L(B)$

First we prove $L(A)L(B) \subseteq L(C)$.

Consider the word $w = w_1w_2$ where $w_1 \in L(A)$ and $w_2 \in L(B)$.

$\Rightarrow h'(g'(w_1, f')) = 1$ and $h''(g''(w_2, f'')) = 1$ where f' and f'' are the characteristic vectors of F' and F''. Start with $h(g(w_1w_2, f))$ where f is the characteristic vector of F and $F = F' \bigcup F'' \Rightarrow f = (f', f'')$. First the machine C reads w_2 and it activates all the sates of C, according to the transition rule $g_q(a, u) = g''_q(a, u)$ if $q \in Q'', a \in \Sigma_2$, the value of f'' changes for the word w_2 and the value of f' remains as it is using

$$g_q(a, u) = \begin{cases} 0 & \text{if } q \notin F', a \in \Sigma_2 \\ 1 & \text{if } q \in F', a \in \Sigma_2 \end{cases}$$

Then the machine C reads the word w_1, the value of f' changes to the transition rule $g_q(a, u) = g'_q(a, u)$ if $q \in Q'$, $a \in \Sigma_1$ and f'' remains as it is using $g_q(a, u) = q$ if $q \in Q''$, $a \in \Sigma_1$.

For the machine $C, h = h' \bigwedge h''$

Therefore

$$\begin{aligned} h(g(w_1w_2, f)) &= h'(g'(w_1, f')) \bigwedge h''(g''(w_2, f'')) \\ &= 1 \bigwedge 1 \\ &= 1 \end{aligned}$$

$\Rightarrow w_1w_2 \in L(C)$

$\Rightarrow w \in L(C)$

Therefore $L(A)L(B) \subseteq L(C)$

To prove $L(C) \subseteq L(A)L(B)$

Let $w \in L(C)$, $w \in (\Sigma_1 \bigcup \Sigma_2)^*$

Let $w = \alpha\beta$, where α is the largest prefix of w such that $\alpha \in \Sigma_1^*$ and $\alpha \neq \varepsilon, \beta \neq \varepsilon$

$w = \alpha\beta \in L(C)$

$\Rightarrow h(g(\alpha\beta, f)) = 1$

$\Rightarrow h(g(\alpha\beta, (f', f''))) = 1$

Since $\alpha \in \Sigma_1^*, h = h' \bigwedge h''$

$\Rightarrow h'(g'(\alpha, f')) \bigwedge h''(g''(\beta, f'')) = 1$

$\Rightarrow h'(g'(\alpha, f')) = 1$ and $h''(g''(\beta, f'')) = 1$

$\Rightarrow \alpha \in L(A)$ and $\beta \in L(B)$

Therefore $w = \alpha\beta \in L(A)L(B)$

$L(C) \subseteq L(A)L(B)$

Therefore $L(C) = L(A)L(B)$ □

3.2 Complementation

Theorem 2. *An h-AFA A accepts the language L with n states then L^c is accepted by an h-AFA B with same number of states.*

Proof: Let $A = (Q, \Sigma, h, F, g)$ be an h-AFA that accepts the language L with n states and $L(A) = \{w \in \Sigma^* / h(g(w, f)) = 1\}$

Then we can construct an h-AFA B with same number of states to accept the language $L^c(A)$ as follows :

Let $B = (Q, \Sigma, h^*, F', g^*)$, where $F' = Q - F$ and g^* and h^* are got by exchanging conjunction and disjunction as in g and h. The other components remain unchanged.

Claim : $L^c(A) = L(B)$

Consider a word $w \in L^c(A)$

That is $w \notin L(A) \Rightarrow h(g(w, f)) = 0$, where f is a characteristic vector of F

By the construction of the machine B, it is clear that for the word w, $h^*(g^*(w, f'))$ is a boolean vector which is got by exchanging conjunction and disjunction as in A, where f' is the characteristic vector of F'.

For B, $F' = Q - F$ $h^* = \overline{h}$, $g^* = \overline{g}$,

$\Rightarrow h^*(g^*(w, f')) = \overline{h(g(w, f))} = \overline{0} = 1$

$\Rightarrow w \in L(B)$

$L^c(A) \subseteq L(B)$

On the other hand, if $w \in L(B)$ then

$h^*(g^*(w, f')) = 1$

$\Rightarrow \overline{h(g(w, f))} = 1$

$\Rightarrow h(g(w, f)) = 0$

$\Rightarrow w \in L^c(A)$

That is $L(B) \subseteq L^c(A)$

Therefore $L(B) = L^c(A)$ □

3.3 Intersection

Theorem 3. *For any positive integers m, n Let A be an m-state h-AFA and B be an n-state h-AFA. Then m + n states are sufficient in the worst case for an h-AFA C to accept the language $L(A) \cap L(B)$.*

Proof: Let A be an h-AFA $A = (Q', \Sigma, h', F', g')$ accepts the language $L(A)$ with m states.

Let B be an h-AFA $B = (Q'', \Sigma, h'', F'', g'')$ accepts the language $L(B)$ with n states and $Q' \bigcap Q'' = \phi$.

We construct an $m + n$ state h-AFA $C = (Q, \Sigma, h, F, g)$ such that $L(C) = L(A) \cap L(B)$ as follows.

$Q = Q' \bigcup Q''$,

$h = h' \bigwedge h''$,

$F = F' \cup F''$

and g is defined as follows.

$$g_q(a, u) = \begin{cases} g'_q(a, u) & \text{if } q \in Q', a \in \Sigma \\ g''_q(a, u) & \text{if } q \in Q'', a \in \Sigma \end{cases}$$

Claim : $L(C) = L(A) \bigcap L(B)$.

First we prove $L(A) \bigcap L(B) \subseteq L(C)$.

Consider a word $w \in L(A) \bigcap L(B)$

Therefore $w \in L(A)$ and $w \in L(B)$.

$\Rightarrow h'(g'(w, f')) = 1$ and $h''(g''(w, f'')) = 1$

where f' and f'' are the characteristic vectors of F' and F''

$\Rightarrow h'(g'(w, f')) \bigwedge h''(g''(w, f'')) = 1$

For the machine C, $F = F' \bigcup F''$ and $h = h' \bigwedge h''$
$\Rightarrow w \in L(C)$
Therefore $L(A)L(B) \subseteq L(C)$
Consider a word $w \in L(C)$
$\Rightarrow h(g(w, f)) = 1$
$\Rightarrow h'(g'(w, f')) \bigwedge h''(g''(w, f'')) = 1$
$\Rightarrow h'(g'(w, f')) = 1$ and $h''(g''(w, f'')) = 1$
$\Rightarrow w \in L(A)$ and $w \in L(B)$
$\Rightarrow w \in L(A) \bigcap L(B)$
Therefore $L(C) \subseteq L(A) \bigcap L(B)$
Therefore $L(C) = L(A) \bigcap L(B)$ □

3.4 Difference

Theorem 4. *For any positive integers m, n Let A be an m-state h-AFA and B be an n-state h-AFA. Then $m + n$ states are sufficient for an h-AFA C to accept the language $L(A) - L(B)$.*

Proof: Let A be an h-AFA $A = (Q', \Sigma, h', F', g')$ accepts the language $L(A)$ with m states.

Let B be an h-AFA $B = (Q'', \Sigma, h'', F'', g'')$ accepts the language $L(B)$ with n states and $Q' \bigcap Q'' = \phi$.

We construct an m+n state h-AFA $C = (Q, \Sigma, h, F, g)$ which accepts $L(A) - L(B)$ as follows.

$$Q = Q' \bigcup Q'',$$
$$h = h' \bigwedge \overline{h''},$$
$$F = F' \bigcup \{Q'' - F''\}$$

and g is defined as follows.

$$g_q(a, u) = \begin{cases} g'_q(a, u) & \text{if } q \in Q', a \in \Sigma \\ \overline{g''_q(a, u)} & \text{if } q \in Q'', a \in \Sigma \end{cases}$$

where $\overline{g}$ is a boolean function as in g in which conjunctions and disjunctions are interchanged.

Claim : $L(C) = L(A) - L(B)$.

Consider a word $w \in L(C)$, $w \in \Sigma^*$

$\Rightarrow h(g(w, f)) = 1$ where f is a characteristic vector of F.

Since $F = F' \bigcup \{Q'' - F''\}$, $f = (f', \overline{f''})$

$h(g(w, f)) = 1$ iff $h(g(w, (f', \overline{f''})) = 1$

$$\text{iff } (h' \bigwedge \overline{h''})(g(w, (f', \overline{f''})) = 1$$
$$\text{iff } h'(g'(w, f')) \bigwedge \overline{h''}(\overline{g''}(w, \overline{f''})) = 1$$
$$\text{iff } h'(g'(w, f')) = 1 \text{ and } \overline{h''}(\overline{g''}(w, \overline{f''})) = 1$$
$$\text{iff } w \in L(A) \text{ and } w \in \overline{L(B)}$$
$$\text{iff } w \in L(A) \bigcap \overline{L(B)}$$

iff $w \in L(A) - L(B)$

Therefore $w \in L(C)$ iff $w \in L(A) - L(B)$
That is $L(C) = L(A) - L(B)$ □

3.5 Shuffle

Given two strings s_1 and s_2 the shuffle of s_1 and s_2 is the set of strings obtained by taking the characters of s_1 and interleaving them with the characters of s_2 so that the relative order of the characters in each string is maintained.

The shuffle of two languages L_1 and L_2 is defined as

$$L1||L_2 = \bigcup_{\substack{x \in L1 \\ y \in L2}} x||y$$

Theorem 5. *Let A be an h-AFA with m states accepts the language L(A) and B be an h-AFA with n states accepts the language L(B) then there exist an h-AFA C with m + n states which accepts the shuffle of two languages L(A) and L(B).*

Proof: Let $A = (Q', \Sigma_1, h', F', g')$ be an h-AFA with m states and $B = (Q'', \Sigma_2, h'', F'', g'')$ be an h-AFA with n states . Then the shuffle product of these two languages is accepted by an h-AFA $C = (Q, \Sigma, h, F, g)$, where

$$Q = Q' \bigcup Q'',$$
$$\Sigma = \Sigma_1 \bigcup \Sigma_2,$$
$$F = F' \bigcup F'',$$
$$h = h' \bigwedge h''$$

and g is defined as

$$g_q(a,u) = \begin{cases} g'_q(a,u) & \text{if } q \in Q', a \in \Sigma_1 \\ g''_q(a,u) & \text{if } q \in Q'', a \in \Sigma_2 \\ q & \text{if } q \in Q', a \in \Sigma_2 \\ q & \text{if } q \in Q'', a \in \Sigma_1 \end{cases}$$

Claim : $L(C) = L(A)||L(B)$
Let $w \in L(C)$, $w \in (\Sigma_1 \bigcup \Sigma_2)^*$
Since $w \in L(C)$ we've $h(g(w, f)) = 1$ where f is the characteristic vector of F.
$(h' \bigwedge h'')(g(w, f)) = 1$
$h'(g(w, f)) = 1$ and $h''(g(w, f)) = 1$
where h' and h'' are the accepting boolean functions for the machine A and B.
Since $g_q(a, u) = q$ for $q \in Q', a \in \Sigma_2$ and $h'(g(w, f)) = 1$
$\Rightarrow g(w, f) = g'(w, f')$, where f' is the characteristic vector of F'.
Therefore $h'(g'(w, f')) = h'(g(w, f)) = 1$.
Since $g_q(a, u) = q$ for $q \in Q'', a \in \Sigma_1$ and $h''(g(w, f)) = 1$
$\Rightarrow g(w, f) = g''(w, f'')$, where f'' is the characteristic vector of F''.
Therefore $h''(g''(w, f'')) = h''(g(w, f)) = 1$

The two values $h'(g'(w, f'))$ and $h''(g''(w, f''))$ are equal to one implies that the relative order of the words in w one is from Σ_1^* and other one is from Σ_2^* are maintained as in $L(A)$ and $L(B)$

$\Rightarrow w \in L(A)||L(B)$

$\Rightarrow L(C) \subseteq L(A)||L(B)$

To prove $L(A)||L(B) \in L(C)$

Let $w \in L(A)||L(B)$, $w = \alpha||\beta, \alpha \in L(A)$ and $\beta \in L(B)$

$\Rightarrow h'(g'(\alpha, f')) = 1$ and $h''(g''(\beta, f'')) = 1$

$\Rightarrow h'(g'(\alpha, f')) \bigwedge h''(g''(\beta, f'')) = 1$

$\Rightarrow h(g(w, f) = 1$

$\Rightarrow w \in L(C)$

Therefore $L(A)||L(B) \subseteq L(C)$.

Therefore $L(C) = L(A)||L(B)$ □

Acknowledgement

The authors would like to express their gratitude to Dr. (Mrs.) R. Rama, Department of Mathematics, I.I.T Madras, for introducing the state complexity problem on regular languages to the authors and for her constant encouragement.

References

[1] A.K. Chandra, D.C. Kozen, L.J. Stockmeyer: Alternation. *JACM* **28** 1981, 114–133.

[2] A. Fellah, H. Jurgensen, S. Yu: Constructions for alternating finite automata. *International journal of computer mathematics* **35** 1990, 117–132.

[3] K. Salomaa, S. Yu, Q. Zhuang: The state complexity of some basic operations on regular languages. *Theoretical computer science* **125** 1994, 315–328.

[4] S. Yu: Regular languages. In : G. Rozenberg, A. Salomaa (Eds), *Handbook of Formal Languages* **1** Springer, Berlin, New york, 1997, 41–110.

[5] C. Campeanu, K. CulikII, K. Salomaa, S. Yu: State complexity of basic operations on finite languages. In : O. Boldt, H. Jurgensen (Eds.), Proc. Fourth Internat. Workshop on Implementing Automata (WIA'99), *Lecture Notes in Computer Science* **2214** Springer, Heidelberg, 2001, 60–70.

[6] G. Pighizzini: Unary language concatenation and its stats complexity. In : S. Yu, A. Paun (Eds.), Implementation and Application of Automata: Fifth Internat. Conference, CIAA 2000, *Lecture Notes in Computer Science* **2088**, Springer, Heidelberg, 2001, 252–262.

[7] S. Yu: State complexity of regular languages. *Journal of automata languages and combinatorics* **6** 2001, 221–234.

[8] J. Hromkovic: Descriptional complexity of finite automata: concepts and open problems. *J.Automat.Lang.Comb.* **7** 2002, 519–531.

[9] Galina Jiraskova: State complexity of some operations on binary regular languages. *Theoretical computer science* **330**, 2005, 287–298.

[10] Markus Holzer, Martin Kutrib: On the descriptional complexity of finite automata with modified acceptance condition. *Theoretical computer science* **330** 2005, 267–285.

Kavitha J. and Sethuraman G.
Department of Mathematics
Anna University
Chennai 600025

Formal Language Aspects of Natural Computing– 2006, pp. 171–178

On the Efficiency of a Variant of P Systems with Mobile Membranes

S.N. Krishna

Abstract

P systems with mobile membranes were introduced in [1]. It was shown that without polarizations, we can still achieve completeness and solve hard problems, by simulating P systems with replicated rewriting. However, we needed local division rules for solving hard problems. In this paper, we look at a variant of this model and show that we can solve *SAT* in polynomial time even when the division rules are global.

1 Introduction

The present paper can be seen as a contribution to the important question whether computationally hard problems can be solved by means of P systems with active membranes without using polarizations. We continue here the approach started in [1], where P systems with the membranes evolving by means of endocytosis and exocytosis operations were considered. Such systems were proved to be computationally universal, and, when provided with the possibility of dividing membranes, also able to solve hard problems in polynomial time. However, the division rules used in [1] were localized (a membrane is divided only when placed in a specified membrane). Here we remove this restriction and still solve *SAT* in polynomial time with respect to the number of clauses and of variables. Note that neither in [1] nor here the membranes have electrical charges, and that the labels of membranes are not changed during the computation.

2 P Systems with Endocytosis and Exocytosis

We refer to [4] for the elements of formal language theory we use here. We only specify that for a string $x \in V^*$ and a symbol $a \in V$, we denote by $|x|$ the length of x and by $|x|_a$ the number of occurrences of the symbol a in the string x. For $w \in V^*$ with $V = \{a_1, \ldots, a_n\}$, we denote by $\Psi_V(w)$ the Parikh vector of w, that is, $\Psi_V(w) = (|w|_{a_1}, \ldots, |w|_{a_n})$; this is extended to languages in a natural way. For a family FL of languages, we denote by $PsFL$ the family of Parikh sets of vectors associated with languages in FL.

A multiset over an alphabet V is represented by a string over V (and by all its permutations) and each string precisely identifies a multiset; the Parikh vector associated with the string indicates the multiplicities of each element of V in the corresponding multiset.

For basic elements of membrane computing we refer to [3]; for the state-of-the art of the domain, the reader may consult the bibliography from the web address `http://psystems.disco.unimib.it`.

We now recall P systems with mobile membranes and introduce the following variant called P systems with mobile membranes having *local evolution* which we investigate below.

A P system with mobile membranes having *local evolution* is a construct

$$\Pi = (V, H, \mu, w_1, \ldots, w_n, R),$$

where:

1. $n \geq 1$ (the initial *degree* of the system);
2. V is an alphabet (its elements are called *objects*);
3. H is a finite set of *labels* for membranes;
4. μ is a *membrane structure*, consisting of n membranes, labeled (not necessarily in a one-to-one manner) with elements of H;
5. $w_1, w_2, \ldots, w_n$ are strings over V, describing the *multisets of objects* placed in the n regions of μ;
6. R is a finite set of *developmental rules*, of the following forms:

   (a) $[_h[_m a \to v]_m]_h$, for $m, h \in H, a \in V, v \in V^*$;
   object evolution rules. These rules are called *local*, because the evolution of object a in m is possible only when m is inside h; if this restriction is not imposed, that is, the evolution of a in m is allowed irrespective of where m is placed, then we say that we have a *global* evolution rule, and write it simply as $[\,_m a \to v]_m$.

   (b) $[_h a]_h[_m\ ]_m \to [_m[_h b]_h]_m$, for $h, m \in H, a, b \in V$;
   endocytosis: an elementary membrane labeled h enters the adjacent membrane labeled m, under the control of object a; the labels h and m remain unchanged during this process, however, the object a may be modified to b during the operation; m is not necessarily an elementary membrane.

   (c) $[_m[_h a]_h]_m \to [_h b]_h[_m\ ]_m$, for $h, m \in H, a, b \in V$;
   exocytosis: an elementary membrane labeled h is sent out of a membrane labeled m, under the control of object a; the labels of the two membranes remain unchanged, but the object a from membrane h may be modified during this operation; membrane m is not necessarily elementary.

   (d) $[_h a]_h \to [_h b]_h[_h c]_h$, for $h \in H, a, b, c \in V$;
   division rules for elementary membranes: in reaction with an object a, the membrane labeled h is divided into two membranes labeled h, with the object a replaced in the two new membranes by possibly new objects.

Note that rules of types (a) in the global version to (d) are like in [1], whereas rules of type (a) in the local version are new. The rules of type (d) as defined above are global; also a local variant can be considered (this was the case in [1]), but we do not need them below.

The rules are applied according to the following principles, most of them customary in membrane computing:

1. All rules are applied in parallel, non-deterministically choosing the rules, the membranes, and the objects, but in such a way that the parallelism is maximal; this means that in each step we apply a set of rules such that no further rule can be added to the set, no further membranes and objects can evolve at the same time.

2. The membrane m from each type (a) – (d) of rules as above is said to be *passive*, while the membrane h is said to be *active*. In any step of a computation, any object and any active membrane can be involved in at most one rule, but the passive membranes are

not considered involved in the use of rules (hence they can be used by several rules at the same time as passive membranes); for instance, a global evolution rule $[\,_m a \rightarrow v]_m$, of type (a), is considered to involve only the object a, not also the membrane m. In the case of the local evolution rules, membrane h is active. This choice is consistent with the fact that we do not use polarizations and we do not change the labels, hence we can use rules in parallel, for the same passive membrane, no conflict is possible (on polarizations or labels).

3. The evolution of objects and membranes takes place in a bottom-up manner. After having a (maximal) set of rules chosen, they are applied starting from the innermost membranes, level by level, up to the skin membrane (all these sub-steps form a unique evolution step, called a *transition* step).

4. When a membrane is moved across another membrane, by endocytosis or exocytosis, its whole contents (its objects) are moved; because of the bottom-up way of using the rules, the inner objects first evolve (if there are rules applicable for them), and then any membrane is moved with the contents as obtained after this inner evolution.

5. If a membrane is divided, then its contents are replicated in the two new copies, with the exception of the object involved in the division rule, which may be changed by new objects in the new membranes. If at the same time a membrane is divided and there are objects in this membrane which evolve by means of a rule of type (a), then in the new copies of the membrane one introduces the result of evolution. Similarly, if a membrane l enters by endocytosis a membrane h which is divided, then in each of the two copies of h we will have a copy of membrane l (when dividing, membrane h is elementary, but the new membranes h are no longer so).

6. If a membrane exits the system (by exocytosis), then its evolution stops, even if there are rules of types (a) or (d) which would be applicable to it provided that the membrane would be in the system.

7. The skin membrane can never be divided.

8. All objects and membranes which do not evolve at a given step (for a given choice of rules which is maximal) are passed unchanged to the next configuration of the system.

Note that we divide and move by endocytosis and exocytosis only elementary membranes.

By using the rules in this way, we get transitions among the configurations of the system. A sequence of transitions is a computation, and a computation is successful if it halts (it reaches a configuration where no rule can be applied). During a computation, membranes can leave the skin membrane (by means of rules of type (c)). The multiplicity vector of the multiset from such a membrane is considered as a result of the computation. Thus, the result of a halting computation consists of all vectors describing the multiplicity of objects from all membranes sent out of the system during the computation; a non-halting computation provides no output. The set of all vectors of natural numbers produced in this way by a system Π is denoted by $Ps(\Pi)$.

We emphasize the difference in defining the result of a computation between our case and the standard one: classically, at the end of a computation one counts all symbol-objects sent out of the system, and their multiplicity vector is the vector provided by the computation; in our case, because we send into the environment membranes, not free objects, each membrane which exits the system gives a vector, hence a computation can produce several vectors, all of them considered in the set $Ps(\Pi)$.

The family of all sets $Ps(\Pi)$ generated by systems using division rules, local evolution rules, endocytosis and exocytosis rules, is denoted by $PsMP(div, levol, endo, exo)$; when global evolution rules are used, we replace *levol* with *gevol*.

If a type of rules is not used, then we omit its "name" from the list.

Convention: When comparing the power of two systems Π, Π', we consider $Ps(\Pi) = Ps(\Pi')$ if and only if $Ps(\Pi) - \{(0, 0, \ldots, 0)\} = Ps(\Pi') - \{(0, 0, \ldots, 0)\}$, that is, the null vector is ignored. This corresponds to the fact that, in many cases in language theory, when comparing two grammars or automata, the empty string is ignored – and this last assumption is also made here.

We use the shorthand notations *endo, exo, div, levol*(*gevol*) for endocytosis rules, exocytosis rules, division and local (global) evolution rules.

By *RE* we denote the family of recursively enumerable languages, hence *PsRE* is the family of all Turing computable sets of vectors of natural numbers.

Theorem 1. $PsMP_4(endo, exo, gevol) \subseteq PsMP_4(endo, exo, levol) = PsRE.$

Proof: It has been proved in [2] that 4 membranes are sufficient for completeness using endo, exo, and global evolution rules. □

3 Solving SAT

The SAT problem (satisfiability of propositional formulae in the conjunctive normal form) is probably the best known **NP**-complete problem. A propositional formula C is in the conjunctive normal form if it is a conjunction of disjunctions, that is,

$$C = C_1 \wedge C_2 \wedge \cdots \wedge C_m,$$

where each C_i, $1 \leq i \leq m$, is a clause of the form

$$C_i = y_{i,1} \vee y_{i,2} \vee \cdots \vee y_{i,p},$$

with each literal $y_{i,j}$ being either a propositional variable, x_s, or its negation, $\neg x_s$.

Theorem 2. *The SAT problem can be solved in linear time with respect to the number of clauses and number of variables by a P system with mobile membranes using rules of the types endo, exo, div, gevol, and levol.*

Proof: Let us consider a propositional formula $C = C_1 \wedge C_2 \wedge \cdots \wedge C_m$, with $C_i = y_{i,1} \vee y_{i,2} \vee \cdots \vee y_{i,p}$, and $y_{i,j} \in \{x_s, \neg x_s \mid 1 \leq s \leq n\}$.

We construct the P system

$$\Pi = (V, H, \mu, w_1, w_2, w_3, w_4, R),$$

where:

$$V = \{a_1, x_2^{i_2}, x_3^{i_3}, \ldots, x_n^{i_n}, d_j \mid 0 \leq i_2 \leq 1, 0 \leq i_3 \leq 2, \ldots, 0 \leq i_n \leq n-1, 0 \leq j \leq n\}$$

$$\cup \{t_j^k, f_j^k \mid 1 \le j \le n, 0 \le k \le n-j\} \cup \{c_r^{jn} \mid 0 \le j \le 2r-2, 1 \le r \le m\}$$
$$\cup \{O, O', O_{i,j}^k, O'^{k}_{i,j}, E_{i,j}^k, E'^{k}_{i,j} \mid 1 \le i, k \le m, 1 \le j \le n\}$$
$$\cup \{b_j, g_j, g'_p \mid 1 \le j \le n, 1 \le p \le n-1\},$$
$$H = \{1, 2, 3, 4\},$$
$$\mu = [_4[_1\ ]_1[_2\ ]_2[_3\ ]_3]_4,$$
$$w_1 = a_1 x_2^0 x_3^0 \ldots x_n^0 d_0,$$
$$w_2 = w_3 = b_1, w_4 = \lambda.$$

The rules are as follows.

1. Generation of the 2^n truth assignments.

$$[_1 a_1]_1 \to [_1 t_1^0]_1 [_1 f_1^0]_1,$$
$$[_1 x_2^0 \to x_2^1, x_3^0 \to x_3^1, \ldots, x_n^0 \to x_n^1, d_0 \to d_1]_1,$$
$$[_2 b_1]_2 \to [_2 b_2]_2 [_2 b_2]_2, [_3 b_1]_3 \to [_3 b_2]_3 [_3 b_2]_3,$$
$$[_1 x_2^1]_1 \to [_1 t_2^0]_1 [_1 f_2^0]_1,$$
$$[_1 x_3^1 \to x_3^2, \ldots, x_n^1 \to x_n^2, d_1 \to d_2, t_1^0 \to t_1^1, f_1^0 \to f_1^1]_1,$$
$$[_2 b_2]_2 \to [_2 b_3]_2 [_2 b_3]_2, [_3 b_2]_3 \to [_3 b_3]_3 [_3 b_3]_3$$
$$\ldots$$
$$[_1 x_n^{n-1}]_1 \to [_1 t_n^0]_1 [_1 f_n^0]_1,$$
$$[_1 d_{n-1} \to d_n, t_1^{n-2} \to t_1^{n-1}, f_1^{n-2} \to f_1^{n-1}]_1,$$
$$[_1 t_2^{n-3} \to t_2^{n-2}, f_2^{n-3} \to f_2^{n-2}]_1,$$
$$\ldots$$
$$[_1 t_{n-1}^0 \to t_{n-1}^1, f_{n-1}^0 \to f_{n-1}^1]_1,$$
$$[_2 b_{n-1}]_2 \to [_2 b_n]_2 [_2 b_n]_2, [_3 b_{n-1}]_3 \to [_3 b_n]_3 [_3 b_n]_3.$$

The object a_1 corresponds to x_1; a_1 divides membrane 1 into two membranes. Simultaneously, the superscripts of all the x_i^j's are incremented. In general, assignments corresponding to variable x_i is made when the superscript is $i-1$. This is continued until all assignments are generated. d_0 is an auxiliary variable which evolves in parallel, and reaches the index value n when all assignments are generated. Clearly, n steps are required for generating all the assignments. By the time 2^n assignments are generated in membrane 1, 2^n copies of membranes 2 and 3 are also created, using the objects b_i, $1 \le i \le n$.

2. In the next step, we look for the clauses satisfied by the truth assignments in each membrane. This is done in parallel across all 2^n membranes labeled 1. In this step, $t_1^{n-1}, t_2^{n-2}, \ldots, t_n^0$ and $f_1^{n-1}, f_2^{n-2}, \ldots, f_n^0$ are rewritten by a string of clauses. If all m clauses can be found in a membrane, then that will be a satisfying assignment for the given CNF.

For $1 \leq j \leq n$,

$$[_1 t_j^{n-j} \rightarrow c_s^{(2s-2)n} c_r^{(2r-2)n} \ldots c_t^{(2t-2)n}]_1,$$

if x_j occurs in clauses $C_r, C_s, \ldots, C_t$,

$$[_1 f_j^{n-j} \rightarrow c_s^{(2s-2)n} c_r^{(2r-2)n} \ldots c_t^{(2t-2)n}]_1,$$

if $\neg x_j$ occurs in clauses $C_r, C_s, \ldots, C_t$,

$$[_1 d_n \rightarrow O]_1, [_2 b_n \rightarrow \lambda]_2, [_3 b_n \rightarrow \lambda]_3.$$

3. Checking if any of the membranes contain m clauses. There may be k occurrences, $0 \leq k \leq n$, of a clause C_j in a membrane. We need to identify only one of them for satisfiability of C_j. The superscripts help in doing this one by one. Since the superscripts are chosen in such a way that they become zero $2n$ steps apart, we can sequentially check for the satisfiability of the clauses $C_1, C_2, \ldots, C_m$ in order. If C_1 is satisfied by any of the membranes 1, then they enter membrane 3, modifying one of the C_1^0's in each into C_1. The object O gets rewritten into O' while membrane 1 enters 3.

$$[_1 c_j^k \rightarrow c_j^{k-1}]_1, k \geq 1, [_1 O \rightarrow O']_1$$
$$[_1 c_j^0]_1 [_2]_2 \rightarrow [_2 [_1 c_j]_1]_2, \ j \text{ even},$$
$$[_1 c_j^0]_1 [_3]_3 \rightarrow [_3 [_1 c_j]_1]_3, \ j \text{ odd}.$$

4. The object O' gets rewritten into $O_{1,1}^1$ while membrane one comes out of 3.

$$[_3 [_1 O']_1]_3 \rightarrow [_1 O_{1,1}^1]_1 [_3]_3$$

5. O' is rewritten to $O_{1,1}^1$ while membrane 1 comes out of 3. In the next step, $O_{1,1}^1$ is rewritten to ${O^1}'_{1,1} g'_1$. Membrane 1 at this point of time may be inside membrane 4 or membrane 3, depending on whether there exists a copy of C_1^0 or not. In either case, ${O^1}'_{1,1}$ gets rewritten to $O_{1,2}^1$. This continues until $O_{1,n}^1$ when we finish verifying if there exists n copies of C_1^0. Now, if C_2^0 exists, then membrane 1 enters membrane 2 and $O_{1,n}^1$ gets rewritten to ${O^1}'_{1,n}$. Further rewriting of ${O^1}'_{1,n}$ is possible only if membrane 1 is inside membrane 4. Otherwise, it gets updated to $E_{2,1}^2$. The indices and superscripts are updated to reflect the current clause, the new counter to keep track of the steps after the switch from O to E, and the number of clauses satisfied. This is continued in all the membranes in parallel, until a $E_{m,1}^m$ or $O_{m,1}^m$ is obtained.

$O_{k,l}^k$ (${O^k}'_{k,l}$) is used to keep track of the number of clauses satisfied so far, and the current clause that is being checked for satisfiability (k). l is used to count the number of steps that have passed ever since a C_j^0 has been identified. If l becomes n in $O_{k,l}^j$ and if membrane 1 is in membrane 4 in the next step, then it means the next clause C_{j+1} is not satisfied (if C_{j+1} were satisfied, there will be at least a c_{j+1}^0 in membrane 1 along with $O_{k,n}^k$, and membrane 1 would be inside membrane 2, $O_{k,n}^k$ is replaced by $O_{k,n}^k$' in any case). The local rule is applied if membrane one is in membrane 4, introducing the trap symbol T. A membrane having T in it remains useless for the rest of the computation. A similar role is played by $E^k_{k,l}$ ($E^{k'}_{k,l}$), the only difference being that when we check satisfiability of a clause with odd index, we use O's and when checking satisfiability of a clause with even index, we use E's. Observe that if C_1 is not satisfied by any of the membranes, then those membranes will be blocked in membrane 2. Similarly, if any clause is not satisfied by an assignment in a membrane, then it will be blocked in membranes 2 or 3 (depending on whether a clause with odd index or even index is not satisfied). The g_i's and g_i''s help eliminate all possible copies of any C_j^0 present in a membrane. The g_i' are rewritten by g when the C_j^0's are exhausted.

For odd k, $1 \leq k \leq m-1$,

$$[_1 O^k_{k,1} \rightarrow O^{k'}_{k,1} g_1']_1,$$
$$[_1 O^k_{k,i} \rightarrow O^{k'}_{k,i}]_1, 2 \leq i \leq n,$$
$$[_1 O^{k'}_{k,i} \rightarrow O^k_{k,i+1}, g_j \rightarrow g_j']_1, 1 \leq i \leq n-1, 2 \leq j \leq n-1,$$
$$[_4 [_1 O^{k'}_{k,n} \rightarrow T]_1]_4,$$
$$[_2 [_1 g_i']_1]_2 \rightarrow [_1 g_{i+1}]_1 [_2]_2, 1 \leq i \leq n-1,$$
$$[_3 [_1 g_i']_1]_3 \rightarrow [_1 g_{i+1}]_1 [_3]_3, 1 \leq i \leq n-1.$$

Similar rules for $E^k_{k,i}$'s.

For even k, $1 \leq k \leq m-1$,

$$[_1 E^k_{k,1} \rightarrow E^{k'}_{k,1} g_1']_1,$$
$$[_1 E^k_{k,i} \rightarrow E^{k'}_{k,i}]_1, 2 \leq i \leq n,$$
$$[_1 E^{k'}_{k,i} \rightarrow E^k_{k,i+1}, g_j \rightarrow g_j']_1, 1 \leq i \leq n-1, 2 \leq j \leq n-1,$$
$$[_4 [_1 E^{k'}_{k,n} \rightarrow T]_1]_4,$$
$$[_2 [_1 g_i']_1]_2 \rightarrow [_1 g_{i+1}]_1 [_2]_2, 1 \leq i \leq n-1,$$
$$[_3 [_1 g_i']_1]_3 \rightarrow [_1 g_{i+1}]_1 [_3]_3, 1 \leq i \leq n-1.$$

Local evolution rules for g_i'.

$$[_4 [_1 g_i' \rightarrow g]_1]_4, 1 \leq i \leq n-1,$$
$$[_2 [_1 O^j_{j,n}]_1]_2 \rightarrow [_1 E^{j+1}_{j+1,1}]_1 [_2]_2, \ j \text{ odd}, 1 \leq j \leq m-1,$$
$$[_3 [_1 E^k_{k,n}]_1]_3 \rightarrow [_1 O^{k+1}_{k+1,1}]_1 [_3]_3, \ k \text{ even}, 1 \leq k \leq m-1.$$

The number of steps taken for each C_j^0, $1 \leq j \leq m-1$, is at most $2n$. Thus, the total number of steps so far is, $n + 1 + 2n(m-1)$.

6. Finally, a membrane is sent out of the system if the the number of clauses satisfied becomes m.

$$[_4 [_1 E^m_{m,1}]_1]_4 \rightarrow [_1 Y]_1 [_4]_4, \quad \text{if } m \text{ even},$$
$$[_4 [_1 O^m_{m,1}]_1]_4 \rightarrow [_1 Y]_1 [_4]_4, \quad \text{if } m \text{ odd}.$$

This requires two steps for generation of an $E^m_{m,1}$ or $O^m_{m,1}$ and one additional step to send out any membrane(s) from the system. Hence, the total number of steps required is $n + 1 + 3 + 2n(m-1) = n(2m-1) + 4$. □

Let us denote by $\mathcal{MMLE}$ the class of P systems with mobile membranes using local evolution rules.

From the above result we have the following theorem:

Theorem 3. $SAT \in PMC_{\mathcal{MMLE}}$.

4 Future Work

We are investigating the possibility of simulating P systems with active membranes using P systems with mobile membranes with or without local evolution. Likewise, we are also trying to estimate the power of local evolution rules.

References

[1] S.N. Krishna, Gh. Păun: P systems with mobile membranes. *Natural Computing*, 4(3) : 255-274 (2005).

[2] S.N. Krishna: The power of mobility: Four membranes suffice. *LNCS Proceedings of CiE 2005*, LNCS 3526: 242-251, Springer, (2005).

[3] Gh. Păun: *Membrane Computing. An Introduction.* Springer-Verlag, Berlin, 2002.

[4] G. Rozenberg, A. Salomaa, eds.: *Handbook of Formal Languages.* Springer-Verlag, Berlin, 1997.

S.N. Krishna
Department of Computer Science and Engineering
Indian Institute of Technology, Bombay
Powai, Mumbai, India
email: krishnas@cse.iitb.ac.in

Formal Language Aspects of Natural Computing– 2006, pp. 179–185

An Infinite Hierarchy for Some Variants of P Systems

S.N. Krishna and R. Rama

Abstract

In this paper, we introduce two special variants of P systems based on sequential rewriting and partially parallel rewriting that allow us to establish an infinite hierarchy with respect to the number of membranes. The key features used in establishing this hierarchy are: the underlying control structure and the shape of the membrane structure used, the determinism of the rules in each membrane and the absence of priorities.

1 Introduction

P Systems were introduced by Gh.Păun in [4] as a class of distributed parallel computing models, inspired by biological systems. The basic ingredients are a *membrane structure*, consisting of several membranes embedded in a main membrane called *skin*, and delimiting regions where certain objects are placed (see Figure 1). The objects placed in a region evolve according to *evolution rules* associated with the region. A *computation* is a sequence of consecutive *configurations* obtained by applying the evolution rules in parallel to the objects in every region in a maximal manner. The *result* of a halting computation is the set of objects present in a specified *output membrane* or the set of objects that are sent out of the system. Many variants of P Systems have been shown to be computationally complete [3, 4, 6, 7, 8].

One of the interesting open problems in the area of P Systems [9] so far has been the question whether suitable variants could be defined that would allow for establishing an infinite hierarchy with respect to the number of membranes. One such hierarchy has been established in [1] using symbol objects and strings with parallel rewriting in the Lindenmayer sense. In this paper we investigate two simple variants of P systems with string objects that allow us to show that the hierarchy with respect to the number of membranes do not collapse. One of these variants uses sequential rewriting as the control structure while the second variant uses partially parallel rewriting.

Our variants have many features in common to the system introduced in [1] viz., the determinism of the rules in each membrane, and the absence of priority relations among the rules, to state a few. The shape of the membrane structure used here is a star-like tree whereas linear structures have been used in [1]. In fact, the results proved in [1] hold good even when star-like trees are used, provided the movement of objects/strings take place using rules of the form $u \rightarrow (v, tar)$. It is worthwhile to examine P systems with restrictions in applying the rules as in [1], having a star-like tree as the underlying membrane structure and having sequential or partially parallel rewriting as the control structure.

Now we shall recall the definitions of sequential rewriting P systems and partially parallel P systems. Sequential rewriting P systems were introduced in [4]. The objects used to encode information in sequential rewriting P systems are strings. In sequential rewriting P systems, each string is processed by exactly one rule during a transition step. If several rules can be applied to a string, maybe in several places each, then we take only one rule and only one possibility to apply it and consider the obtained string as the next state of the object described by the string.

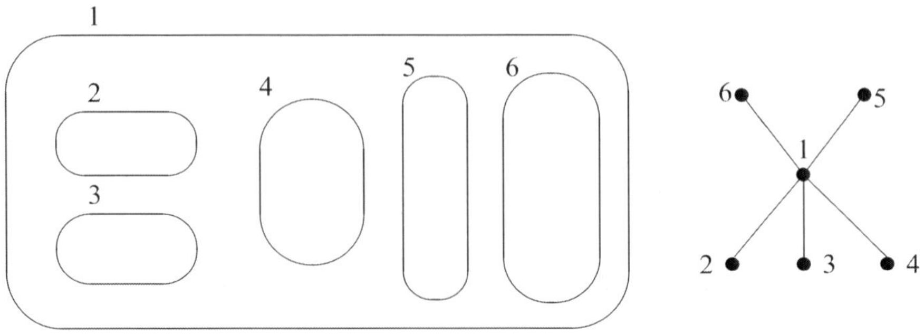

Figure 1. Membrane structure $[_1\ [_2\ ]_2\ [_3\ ]_3\ [_4\ ]_4\ [_5\ ]_5\ [_6\ ]_6\ ]_1$.

Partially parallel P systems were introduced in [3]. What is specific to this system is the way of rewriting the strings and to pass them through membranes. Specifically, each string is rewritten in a partially parallel manner: take a string w in a region i; for each symbol $a \in V$ which appears in w and for which there are rules of the form $a \to (u, tar)$ in the set R_i, we have to rewrite exactly one occurrence (nondeterministically chosen) of a in w (by a rule nondeterministically chosen). Assume that we rewrite in this way n symbols, by rules which have associated targets of the form $here, out, in_j$ of several types and that there are n_{tar} targets of each type. Then, the string obtained by rewriting is sent to the membrane indicated by the target with the maximal n_{tar} (the target which appears the maximal number of times in the used rules). As usual, when several targets have the same maximal number of occurrences, then one of them is nondeterministically chosen.

The work of both these kinds of P systems are defined as usual in the P systems area: in each time unit, each string which can be rewritten is rewritten; the strings and the rules are localized to regions, so the strings in a given region are rewritten only by rules in that region; starting from a given configuration, we pass to another configuration; a sequence of transitions form a computation, and we consider as successful computations only the halting ones (the computations which reach a configuration where no rule can be applied); the result of a halting computation consists of either all strings which are present in a specified output membrane or of all strings that are sent out of the system during the computation.

In the following two sections, we define special variants of sequential and partially parallel rewriting P systems respectively, and also construct suitable languages that allow us to establish infinite hierarchies.

2 Special Variants of P Systems with Sequential Rewriting

In this section, we consider the following special variant of sequential rewriting P systems:

(1) The underlying membrane structure is a star-like tree $[_0\ [_1\ ]_1 \ldots [_d\ ]_d\ ]_0$;

(2) The only axiom must be in the skin membrane and this axiom will be listed in $L(\Pi)$;

(3) Each symbol can have atmost one rule in a membrane.

(4) We denote by V the basic alphabet and by A the set of symbols forming the axiom. Then, a rewriting rule in any membrane i must be of the form $a \to (w, tar)$ where $a \in V$, $w \in A^*\{V - A\}$ or $w \in \{V - A\}A^*$ or $w \in A^+$, and $tar \in \{here, out, in_j\}$ where

- *here* means that the string remains in the same membrane after it is rewritten;
- *out* means that the string is sent out to the adjacent membrane after it is rewritten;
- in_j means that the string is sent to a membrane j, adjacent to i after it is rewritten.

The form of the rules is such that an element of V can be rewritten into a string containing atmost one symbol from $V - A$.

(5) The result of a halting computation is the string finally appearing in a specified output membrane other than the skin.

This special variant of sequential rewriting P Systems are inspired by the following variant of the traditional *classroom model* of problem solving: there is a unique leader(skin) and pupils (each inner membrane). The leader alone can communicate with each pupil, no communication between the pupils are allowed. The leader has the discretion to choose any pupil for working on a problem. The chosen pupil is supposed to work on the problem using the data supplied by the leader (a problem over A) and usage of other external resources $(V - A)$ should be minimal. This pupil should communicate to the leader the outcome of his work if he has not completed solving the problem. The leader will then choose some other pupil (or may request the same fellow to try again) to continue the work after examining the outcome and making necessary changes if required. The problem will be solved ultimately by some pupil and then no more communications with the leader will be made since no more efforts are required.

We shall describe a sequential rewriting P system obeying the above constraints in the following way: $\Pi = (V, d, w, R_0, R_1, \ldots, R_{d-1})$ where V is the basic alphabet, d is the number of membranes in Π, w is the axiom over V, and $R_i, 0 \leq i \leq d-1$ are the sets of rules governing the activities of the membranes i. We shall denote by $SRP(d, *, in)$ the family of languages generated by this special variant of sequential rewriting P systems(d stands for the maximum number of membranes used, $*$ stands for the membrane structure and *in* stands for reading the output in a specified inner membrane).

Theorem 1. $SRP(d, *, in) \supset SRP(d-1, *, in), d \geq 1.$

Proof: The inclusion follows directly from the definition.

The strictness of the inclusion is established by the languages

$$L_d = \{a_1^{n_1} a_2^{n_2} a_3^{n_3} \ldots a_{d-2}^{n_{d-2}} a_{d-1}^{n_{d-1}} a_d^{n_d} \mid n_i = n_{i-1} + n_{i-2}, 3 \leq i \leq d, 1 \leq n_1, n_2, \ldots, n_d\}$$

First we show that L_d can be generated by the $SRP(d, *, in)$ system

$$\Pi = (V, d, w, R_0, R_1, \ldots, R_{d-1})$$

where

$$V = \{a_1, \ldots, a_d, a_1', \ldots, a_d', b_2, \ldots, b_{d-2}\},$$
$$w = a_1 a_2 a_3^2 a_4^3 \ldots a_{d-2}^{n_{d-2}} a_{d-1}^{n_{d-3}+n_{d-2}} a_d^{n_{d-2}+n_{d-1}}, \text{ the smallest word in } L_d,$$

$$
\begin{aligned}
A &= \{a_1, \ldots, a_d\}, \\
R_0 &= \{a_1 \to (a'_1 a_1, here), a_2 \to (a'_2 a_2, here), a_d \to (a_d, in_{d-1})\} \\
&\cup \{a'_1 \to (a_1, in_1), a'_{d-1} \to (a_{d-1}, in_{d-2}), a'_d \to (a_d, in_{d-1})\} \\
&\cup \{a'_i \to (b_i, in_{i-1}), b_i \to (a_i, in_i) \mid 2 \leq i \leq d-2\}, \\
R_i &= \{a_{i+2} \to (a'_{i+2} a_{i+2}, out)\}, 1 \leq i \leq d-2, \\
R_{d-1} &= \{a'_i \to (a'_i, out) \mid i < d-1\} \\
&\cup \{b_i \to (b_i, out) \mid 2 \leq i \leq d-2\} \\
&\cup \{a'_d \to (a_d, here)\}
\end{aligned}
$$

The only inner membrane where computations can halt is $d-1$. Hence, $d-1$ is the output membrane, and all strings which get collected in $d-1$ will belong to $L(\Pi)$. We now prove that the strings getting collected in $d-1$ belong to L_d.

In the initial configuration, we have the smallest word $w = a_1 a_2 a_3^2 a_4^3 \ldots a_{d-2}^x a_{d-1}^y a_d^{x+y}$ in the skin. Applying the rule for a_d, this string can be sent to the output membrane. Otherwise, applying rules to a_1 or a_2, we generate arbitrarily many numbers of a'_1 and a'_2. These are used in the generation of a'_3 in membrane one. Proceeding in this manner, we use the symbols a'_{i-2} and a'_{i-1} for generating a'_i. Corresponding to each occurrence of a'_{i-2} or a'_{i-1}, a symbol a'_i is produced. This ensures that $\#_{a_i} w = \#_{a_{i-1}} w + \#_{a_{i-2}} w$. The computation stops when a string over $\{a_1, \ldots, a_{d-1}, a'_d, a_d\}$ enters membrane $d-1$.

Now we show that L_d cannot be generated by a $SRP(d-1, *, in)$ system. On the contrary, assume that L_d can be generated by a $SRP(k, *, in)$ system $\Pi = (V, k, w, R_0, \ldots, R_{k-1})$ where $k \leq d-1$. As the axiom always is in $L(\Pi)$, the axiom must be $w = a_1 a_2 a_3^2 a_4^3 \ldots a_{d-2}^x a_{d-1}^y a_d^{x+y}$: This is the smallest word in L_d and the rules all are propagating, hence the application of a rule will yield a word of length greater than or equal to the smallest word of L_d, which therefore must be the axiom.

As in $SRP(d, *, in)$ we can have rules for a_1, a_2 in the skin membrane. Rules for incrementing the other a_i's cannot be given in the skin since they are dependent on the previous symbols. Assume that we increment the other symbols in some other membrane, say 1. We shall try to increment the cardinality of more than one symbol viz., a_3 in membrane one. Then we can have a rule $a_3 \to (a'_3 a_3 a_4, out)$ in membrane one which will give $\#_{a_3} w = \#_{a_4} w = \#_{a_1} w + \#_{a_2} w$. After this, the number of a_4's has to be incremented $\#_{a_2} w$ times. The rule for this purpose, if given in membrane one will not work, since there is no priority relation between this rule and the rule $a_3 \to (a'_3 a_3 a_4, out)$. Hence, the rule for incrementing a_4's further have to be given in a separate membrane. This means that for incrementing $a_3, a_4, \ldots, a_d$ we need at least $d-2$ membranes. Since the computation will not stop in any of these membranes, we need one more inner membrane as the output membrane. This means that we need at least $d-1$ inner membranes for generating L_d. In conclusion, L_d cannot be generated by an $SRP(k, *, in)$ system $\Pi = (V, k, w, R_0, \ldots, R_{k-1})$ with $k < d$. □

3 Special Variants of P Systems with Partially Parallel Rewriting

Consider the following very special variant of sequential rewriting P systems, where:

(1) The underlying membrane structure is a star-like tree $[_0 [_1]_1 \ldots [_d]_d]_0$;

(2) The only axiom must be in the skin membrane and this axiom will be listed in $L(\Pi)$;

(3) Each symbol can have atmost one rule in a membrane. No target membranes will be specified for any rule. The string obtained after rewriting can move to any of the adjacent membranes. This means that a target of $\{here, out, in_j\}$ can be assumed while applying each rule and the resultant string can be moved accordingly;

(4) We denote by V the basic alphabet and by A the set of symbols forming the axiom. Then, any rewriting rule must be of the form $a \to w$, $a \in V$, $w \in A^+$;

(5) The result of a halting computation is the string that is sent out of the system.

This variant of partially parallel rewriting P systems are also inspired by the variant of the *classroom model* described above: the leader becomes more strict in this case; since each pupil is capable of working faster than the old set of pupils (partially parallel rewriting in contrast to sequential rewriting), they are not allowed to use any extra resources other than the ones supplied by the leader.

We shall describe a partially parallel rewriting P system obeying the above constraints in the following way: $\Pi = (V, d, w, R_0, R_1, \ldots, R_{d-1})$ where V is the basic alphabet, d is the number of membranes in Π, w is the axiom over V, and $R_i, 0 \le i \le d-1$ are the sets of rules governing the activities of membrane i. We shall denote by $PPP(d, *, \infty)$ the family of languages generated by this special variant of partially parallel rewriting P systems(d stands for the number of membranes used, $*$ stands for the membrane structure and ∞ represents that the output is read outside the system).

Theorem 2. $PPP(d+1, *, \infty) \supset PPP(d, *, \infty), d \ge 1.$

Proof: We first show the inclusion $PPP(d+1, *, \infty) \supseteq PPP(d, *, \infty)$. For any $PPP(d, *, \infty)$ system $\Pi = (V, d, w, R_0, R_1, \ldots, R_{d-1})$ we can construct a $PPP(d+1, *, \infty)$ system $\Pi = (V, d+1, w, R_0, R_1, \ldots, R_d)$ where R_d has no rules.

The strictness of the inclusion is established by the languages

$$L_d = \{a_1^{n_1} a_2^{n_2} \ldots a_d^{n_d} \mid 1 \le n_1 \le n_2 \le \ldots \le n_d\}$$

First we show that L_d can be generated by the $PPP(d, *, \infty)$ system

$$\Pi = (V, d+1, w, R_0, R_1, \ldots, R_d)$$

where

$$\begin{aligned} V = A &= \{a_1, a_2, \ldots, a_d\}, \\ w &= a_1 a_2, \ldots a_d, \\ R_0 &= \{a_i \to a_i \mid 1 \le i \le d\} \\ R_i &= \{a_j \to a_j \mid j \le i\} \cup \{a_l \to a_l^2 \mid l \ge i\}, 1 \le i \le d \end{aligned}$$

In the initial configuration, we have the axiom $a_1 a_2 \ldots a_d$, the smallest word in L_d in the skin membrane. This string itself can be sent out of the system. This string can also move to any of the membranes i, $1 \le i \le d$ depending on the target that has occurred the maximum number of times. If the string moves to a membrane $j < d$, the symbols a_k, $k >= j$ alone will be incremented. This process of incrementing symbols whose indexes are greater than i inside membranes $j <= i$ is continued and ultimately, we obtain a string of the specified form, which can be sent out of the system.

Now we show that L_d cannot be generated by a *PPP*$(d, *, \infty)$ system. On the contrary, assume that L_d can be generated by a *PPP*$(k, *, \infty)$ system where $k \leq d$. Since the smallest word in the language is $a_1 a_2 \dots a_d$, and since all the rules are propagating, this must be the axiom. If we have rules that increase the number of occurrences of each a_i in the skin membrane itself, we cannot obtain words where the number of occurrences of some of the leading symbols are 1. Hence, we should have such rules inside a specific membrane. When symbols of a particular index i are incremented in a region, then there should also be rules for incrementing symbols whose indexes are higher than i in the same region.

Consider an instance of the language where the number of occurrences of each a_i is more than one. Assume that we have in some membrane k inside the skin membrane incrementing rules for all the a_i. After a few steps, we would have equal number of all the symbols. To increment the symbols a_i, $i > 1$ further, we need to move to another membrane. Assume that we can remain in the same membrane for this purpose. Then, we need to do some renaming of a_1 in a particular step, or should have rules of the form $a_1 \rightarrow a_1, a_1 \rightarrow a'_1$, both of which are not permissible due to the constraints on the system. Since we have d symbols, we need to have at least d inner regions. In conclusion, L_d cannot be generated by a *PPP*$(k, *, \infty)$ system where $k \leq d - 1$. □

4 Final Remarks

We have established infinite hierarchies on two simple variants of P systems using string objects. We would like to emphasize that the main feature that allowed us to establish infinite hierarchies on the special variants of P systems investigated here was the star-like membrane structure, the absence of priorities and the determinism of the rules applied in each region. We conjecture that the infinite hierarchy on *SRP* systems holds good even in the absence of constraint 4. Similarly, we conjecture that the hierarchy for *PPP* system holds good in the absence of constraint 3.

References

[1] Rudolf Freund, Special Variants of P Systems Inducing an Infinite Hierarchy with Respect to the Number of Membranes, *Bulletin of EATCS*, **75**, (2001), 209–219.

[2] Oscar H. Ibarra, On membrane hierarchy in P systems, *Theoretical Computer Science*, **334(1-3)**, 2005, 115–129.

[3] S. N.Krishna and R.Rama, A Note on Parallel Rewriting in P Systems, *Bulletin of EATCS*, **73**, (2001), 147–151.

[4] Gh.Păun, Computing with membranes, *Journal of Computer and System Sciences*, **61**, 1, (2000), 108–143.

[5] Gh.Păun, Membrane Computing. An Introduction, Springer-Verlag, Berlin, 2002.

[6] Gh.Păun, P systems with active membranes: Attacking NP complete problems, *Journal of Automata, Languages and Combinatorics*, **6**, (2001), 75-90.

[7] Gh.Păun, G.Rozenberg and A.Salomaa, Membrane computing with external output, *Fundamenta Informaticae*, **41**, (2000), 259–266.

[8] Gh.Păun, Y.Sakakibara and T.Yokomori, P systems on graphs of restricted forms, *Proceedings of IFIP Conference on TCS: Exploring New Frontiers of Theoretical Informatics*, (2000), Sendai, Japan.

[9] `http://psystems.disco.unimib.it` *The P Systems Web Page*

S.N. Krishna
Department of Computer Science & Engg.,
Indian Institute of Technology, Bombay,
Powai, Mumbai-76, India.
`email: krishnas@iitb.ac.in`

R. Rama
Department of Mathematics,
Indian Institute of Technology, Madras,
Chennai-36, Tamilnadu, India.
`email: ramar@iitm.ac.in`

Formal Language Aspects of Natural Computing– 2006, pp. 187–196

Tissue P Systems with Leftmost Derivation

Madhu Mutyam and Kamala Krithivasan

Abstract

In this paper we discuss computational capacity of *tissue P systems with leftmost derivation*. With leftmost rewriting restriction any string present in a cell is rewritten in the leftmost position which can be rewritten by a rule from that cell. We show that systems with four states are computationally universal.

Key words : Rewriting P systems, tissue P systems, rewriting tissue P systems, computational universality, leftmost rewriting, matrix grammars, context-free state grammars.

1 Introduction

Membrane computing [7], initiated by Gh. Păun and hence we call it as *P systems*, is an extensively studied research field in the theoretical computer science stream. Several variants of P systems have been defined and studied thoroughly. Interested reader can refer to [1, 8] for more details about P systems.

In all the basic variants of P systems, membrane structure is represented in a tree-like structure. But it can also be represented it in a graph-like structure. This is first considered in *tissue P systems*, introduced in [6]. Tissue P systems consist of *cells* which are connected in a graph-like manner. Each cell can process symbol objects present in it using its state information and evolution rules.

A variant of tissue P systems, namely, *rewriting tissue P systems*, is proposed in [4] by considering string objects instead of symbol objects. Here context-free rules are used for processing string objects. Using this variant, it is proved that the Satisfiability and the Hamiltonian path problems can be solved in linear time (excluding the time required to construct the tissue P system). It is also proved that such systems having *two* cells and *four* states are computationally universal.

By considering *leftmost rewriting* restriction on the application of rules of rewriting tissue P systems, a variant, namely, *leftmost rewriting tissue P systems*, is proposed in [9]. With leftmost rewriting restriction any string present in a cell is rewritten in the leftmost position which can be rewritten by a rule from that cell. It is proved that leftmost rewriting tissue P systems having either *one* cell and *six* states or *two* cells and *five* states are computationally universal.

The main motivation of this work comes from the following observation. The results presented in [9] are inferior to the results of rewriting tissue P systems obtained in [4]. The reason is that in [4] systems without leftmost rewriting are considered and obtained better results in terms of the number of cells and the number of states. This gives us a wrong impression that leftmost rewriting does not add any extra computational power to the systems under consideration. But the basic objective of using leftmost rewriting restriction in any system of formal language theory is to obtain better results than the one obtained in the systems without leftmost rewriting. For example, context free matrix grammars with appearance checking and λ rules generate the family of recursively enumerable languages whereas matrix grammars with leftmost restriction on rewriting with context free and also λ rules but without any appearance checking generate the same family.

In this paper we consider leftmost rewriting tissue P systems (henceforth, we call them as *tissue P systems with leftmost derivation*) and show that leftmost rewriting restriction adds computational power (in the sense of achieving universality with minimal number of cells and states) to the rewriting tissue P systems. With the help of leftmost derivation, we show that rewriting tissue P systems having $(c, s) \in \{(1,4), (2,2), (2,3), (3,2), (4,1)\}$, where c is the number of cells in the system and s is the maximum number of states in any cell of the system, are computationally universal.

2 Prerequisites

In order to prove universality results we use *matrix grammars* [2] with appearance checking in *strong binary normal form* [3]. Henceforth, unless otherwise stated, we denote by $G = (N, T, S, M, F)$ a matrix grammar with appearance checking in strong binary normal form and with $ac(G) = 2$. The nonterminal alphabet N is of the form $N = N_1 \cup N_2 \cup \{S, \dagger\}$. Let $B^{(1)}$ and $B^{(2)}$ be the two symbols in N_2 for which we have rules $B^{(j)} \to \dagger$, $j \in \{1, 2\}$, in the matrices of M. Let us assume that we have

- k matrices of the form $m_i : (X \to \alpha, A \to x)$, $X \in N_1$, $A \in N_2$, $\alpha \in N_1 \cup \{\lambda\}$, and $x \in (N_2 \cup T)^*$, $1 \leq i \leq k$;
- h matrices of the form $m'_i : (X \to Y, B^{(j)} \to \dagger)$, $X, Y \in N_1$, $j \in \{1, 2\}$.

A context-free *state grammar* [2] is a construct $G = (K, N, T, P, q_0, S)$ where K is a finite set of states, $q_0 \in K$ is the initial state, N, T, and S are defined as in a context-free grammar, and the rules of P are of the form $(p, X) \to (q, w)$ where $p, q \in K$, $X \in N$, $w \in (N \cup T)^*$. We say that $(p, u) \in K \times (N \cup T)^*$ directly derives $(q, v) \in K \times (N \cup T)^*$, iff $u = u_1 X u_2$, $v = u_1 w u_2$, and $(p, X) \to (q, w) \in P$ for some $u_1, u_2 \in (N \cup T)^*$. Then we write $(p, u) \Rightarrow (q, v)$.

Let $\Rightarrow^*$ be the reflexive and transitive closure of $\Rightarrow$. We then define the language generated by G as

$$L(G) = \{y \mid y \in T^*, (q_0, S) \Rightarrow^* (p, y) \text{ for some } p \in K\}.$$

Let $\mathcal{L}(S, X)$ be the family of languages generated by state grammars of type X and $\mathcal{L}(M, X)$ be the family of languages generated by matrix grammars of type X. From [2] we know that for $X \in \{CF, CF - \lambda\}$, $\mathcal{L}(S, X) = \mathcal{L}(M, X)$.

We now recall the definition of rewriting tissue P systems given in [4]. A rewriting tissue P system, of degree m, $m \geq 1$, is a construct

$$\Pi = (V, T, \sigma_1, \sigma_2, \ldots, \sigma_m, syn, i_{out}),$$

where:

- V is an alphabet;
- $T \subseteq V$ is the terminal alphabet;
- $syn \subseteq \{1, \ldots, m\} \times \{1, \ldots, m\}$ (synapses among cells);
- $i_{out} \in \{1, \ldots, m\}$ indicates the output cell;
- σ_i, $1 \leq i \leq m$, are cells of the form

$$\sigma_i = (Q_i, s_{i,0}, L_{i,0}, P_i),$$

where:

- Q_i is a finite set of states;
- $s_{i,0} \in Q_i$ is the initial state;
- $L_{i,0}$ is a finite set of strings over V^* presented initially;
- P_i is a finite set of rules of the form $sX \to s'w(tar)$, where $s, s' \in Q_i, X \in (V-T)$, $w \in V^*$, $tar \in \{here, go\}$ for $i \neq i_{out}$, and $tar \in \{here, go, out\}$ for $i = i_{out}$. When presenting the rules, the indication *here* is in general omitted.

The system can work in any one of the three processing modes, i.e., *min*, *par*, and *max*, for processing strings present in the system. Processing at string-level takes place in a sequential manner, i.e., at any point of time, only one symbol from a string is processed. In *min* mode, one string per cell is processed at a time. In other words, we choose one string at random from all available strings in a cell and process the string using an applicable rule from the cell. Application of the rule results in replacing only one occurrence of the symbol (the left-hand side member of the rule) with the right-hand side member of the rule and changing the state of the cell. In a formal way, *min* mode is defined as follows[4]:

$$sL_i \Rightarrow_{min} s'L'_i \quad \text{iff} \quad \text{there is } x \in L_i \text{ and } sX \to s'w(tar) \in P_i \text{ such that}$$
$$x = x'Xx'' \text{ and } L'_i = (L_i - \{x\}) \cup \{x'wx''\}.$$

In *par* mode, at a time one rule per cell is used for processing all those strings from the cell which can be processed by the rule. In *max* mode, a set of rules, which use the current state of the cell and introduce the same new state after processing strings, is used for processing all those strings which can be processed by the set of rules such that no two strings can be processed by the same rule from the set. For formal definitions of *par* and *max* modes one can refer to [4].

Based on the *tar* value, processed strings can be remained in the same cell (if $tar = here$) or sent to other cells (if $tar = go$) or sent to the environment (if $tar = out$). Rules with $tar = out$ can be applied only in the output cell, whereas rules with other *tar* values can be applied in any cell. Since a cell can be connected to many other cells in the system, three transmission modes, i.e., *repl*, *one*, and *spread*, are defined for sending strings processed by using rules with $tar = go$ from one cell to other cells. In *repl* mode, a processed string in a cell σ_i is sent to all the connected cells of σ_i. In *one* mode, a processed string in a cell σ_i is sent to one of the connected cells of σ_i. In *spread* mode, a processed string in a cell σ_i is sent to a subset of the set of connected cells of σ_i.

Any m-tuple of the form $(s_1L_1, \ldots, s_mL_m)$, with $s_i \in Q_i$ and $L_i \in V^*$, $1 \leq i \leq m$, is called a *configuration* of Π; thus, $(s_{1,0}L_{1,0}, \ldots, s_{m,0}L_{m,0})$ is the *initial configuration* of Π.

We can change the configuration of the system by using rules from the sets P_i, $1 \leq i \leq m$. For a language $L_i \in V^*$, we denote by $L_{i,no}$ the set of strings which are not processed by any rule from P_i, $L_{i,here}$ the set of strings which are processed by rules with $tar = here$ from P_i, $L_{i,go}$ the set of strings which are processed by rules with $tar = go$ from P_i, and $L_{i,out}$ the set of strings which are processed by rules with $tar = out$ from P_i. If cell i is not the output cell, then $L_{i,out} = \emptyset$. Furthermore, for a node i in the graph defined by *syn* we denote the ancestor of node i as $anc(i) = \{j \mid (j, i) \in syn\}$ and the successor of node i as $suc(i) = \{j \mid (i, j) \in syn\}$.

We now define the transition among the configurations of the system Π working in (min, one) mode pair. For other mode pairs one can refer to [4].

Let $C_1 = (s_1L_1, \ldots, s_mL_m)$ and $C_2 = (s'_1L''_1, \ldots, s'_mL''_m)$ be any two configurations of Π. We then write $C_1 \Rightarrow_{min,one} C_2$, if there are languages $L'_1, \ldots, L'_m$ over V such that

$$s_iL_i \Rightarrow_{min} s'_iL'_i,\ 1 \le i \le m, \text{ where each } L'_i = L_{i,no} \cup L_{i,here} \cup L_{i,go} \cup L_{i,out}$$

and

$$L''_i = L_{i,no} \cup L_{i,here} \cup \bigcup_{j \in I_i} L_{j,go},$$

where $I_i \subseteq anc(i)$ such that the set $anc(i)$ was partitioned into $I_1, \ldots, I_m$; at this transition, all non-empty sets of strings of the form $\bigcup_{j\in I_k} L_{j,go}$, $1 \le k \le m$, should be sent to receiving cells (added to L''_l, $1 \le l \le m$).

Note that in the case of the output cell we also remove from L'_i all strings from the set $L_{i,out}$.

If each cell of the system has at most one string, all the three processing modes are same. Similarly, if the system has at most two cells or every cell has at most one successor, all the three transmission modes are same.

As usual a sequence of transitions among configurations of the rewriting tissue P system Π is called a *computation* of Π. A computation is said to be successful if it ends in an *halting configuration*, i.e., a configuration where no rule in any cell of the system can be used. The output of a computation is the set of all strings which are ejected from the output cell to the environment. But we consider the output of a successful computation. The set of all strings over T from the output of a successful computation is said to be the language generated by the rewriting tissue P system.

In tissue P systems with leftmost derivation [9], any string present in a cell is rewritten in the leftmost position which can be rewritten by a rule from that cell. In order to apply a rule, we examine the symbols of a string, step by step, from left to right and the first one which can be rewritten by a rule from the cell of the string is rewritten. If there are several rules with same left-hand side symbol, we can select one of them nondeterministically. We denote by $RtP_{n,m}(\alpha, \beta, left)$ the family of all languages $L(\Pi)$ generated by tissue P systems with leftmost derivation, where n indicates the total number of cells in the system and m indicates the maximum number of states used in any cell of the system, $\alpha \in \{min, max, par\}$ is the processing mode, and $\beta \in \{repl, one, spread\}$ is the transmission mode. When one of the parameters m and n is not bounded, replace it with $*$.

3 Computational Universality

We give universality results for tissue P systems with leftmost derivation working in (min, one) mode with different combinations of number of cells and number of states. Since in all the proofs, we make use of at most one string in a cell and at most one successor for any cell, our results are still valid for systems working in (α, β) mode, where $\alpha \in \{max, par\}$ and $\beta \in \{repl, spread\}$. Interestingly, we can make a point that in all these universality results, the total number of states used in the system is 4. That is, if n_i is the number states in a cell σ_i and m is the number of cells in the system, then $\Sigma_{i=1}^m n_i = 4$.

Theorem 1. $RE = RtP_{1,4}(min, one, left)$.

Proof: We prove only the inclusion $RE \subseteq RtP_{1,4}(min, one, left)$, the reverse inclusion can be obtained using Church-Turing hypothesis. We start by considering a matrix grammar

$G = (N, T, S, M, F)$ in strong binary normal form as given in Section 2, and construct the tissue P system with leftmost derivation

$$\Pi = (V, T, \sigma_1, \emptyset, 1),$$

where

- $V = T \cup N_1 \cup N_2 \cup \{X_i \mid X \in N_1, 1 \le i \le k+1\} \cup \{A_i \mid A \in N_2, 1 \le i \le k+1\} \cup \{A' \mid A \in N_1 \cup N_2\} \cup \{E, Z\}$;
- $\sigma_1 = (\{s, s_1, s^{(1)}, s^{(2)}\}, s, \{XAE\}, P_1)$, where P_1 contains the following rules:
    1. $sX \to s_1X_1, X \in N_1$;
    2. $s_1A \to s_1A', A \in N_2$;
    3. $s_1A \to sA_1, A \in N_2$;
    4. $sX_i \to s_1X_{i+1}, X \in N_1, 1 \le i \le (k-1)$;
    5. $sX_{k+1} \to sZ(out), X \in N_1$;
    6. $s_1A_i \to sA_{i+1}, A \in N_2, 1 \le i \le (k-1)$;
    7. $sX_i \to sY'$, for type 2 matrix $m_i : (X \to Y, A \to x)$;
    8. $sX_i \to s\lambda$, for type 4 matrix $m_i : (X \to \lambda, A \to x)$;
    9. $sA' \to sA, A \in N_2$;
    10. $sA_i \to s_1x$, for type 2 matrix $m_i : (X \to Y, A \to x)$;
    11. $sA_i \to sx$, for type 4 matrix $m_i : (X \to \lambda, A \to x)$;
    12. $s_1X' \to sX, X \in N_1$;
    13. $sE \to s\lambda(out)$;
    14. $s_1E \to s_1Z(out)$;
    15. $sX \to s^{(j)}Y$, for $m_i' : (X \to Y, B^{(j)} \to \dagger), j \in \{1, 2\}$;
    16. $s^{(1)}B^{(1)} \to s^{(1)}Z(out)$;
    17. $s^{(1)}E \to sE$;
    18. $s^{(2)}B^{(2)} \to s^{(2)}Z(out)$;
    19. $s^{(2)}E \to sE$.

The system works as follows:

Initial configuration of the system is $(s\{XAE\})$. Assume that at a particular instant the configuration of the system is $(s\{Xw_1Aw_2E\})$, where $w_1, w_2 \in (N_2 \cup T)^*$.

To simulate a matrix $m_i : (X \to Y, A \to x)$ of type 2, we apply a rule of type $P_1(1)$ so that the configuration of the system becomes $(s_1\{X_1w_1Aw_2E\})$. Now in state s_1, we can apply either a rule of type $P_1(2)$ or a rule of type $P_1(3)$. If we apply rules of type $P_1(2)$ on all nonterminals from the set N_2 present in $X_1w_1Aw_2E$, a trap symbol Z is introduced by the application of the rule $P_1(14)$ (this is the only rule that can be applied on the resulting string). So, in order not to get a trap symbol, we must apply a rule of type $P_1(3)$. Application of a rule of type $P_1(3)$ changes the configuration of the system to $(s\{X_1w_1'A_1w_2E\})$, where w_1' represents a string in which every nonterminal B is replaced with B'. By applying a rule of type $P_1(4)$, we change the configuration of the system to $(s_1\{X_2w_1'A_1w_2E\})$. We now apply the only applicable rule (i.e., of type $P_1(6)$) on the string so that the configuration becomes $(s\{X_2w'_1A_2w_2E\})$. We repeat the process of incrementing the subscripts of both X and A. Let us suppose that at a particular instant the configuration of the system be $(s\{X_iw'_1A_iw_2E\})$. We now apply a rule of type $P_1(7)$ so that X_i is replaced with Y', where $X \rightarrow Y \in m_i$, and thus the configuration becomes $(s\{Y'w'_1A_iw_2E\})$. Since the cell is in state s and there is no nonterminal from the set N_1 present in the string, we can apply rules of type $P_1(9)$ on all applicable nonterminals of the form A'. After applying rules of type $P_1(9)$ on all applicable nonterminals, the configuration becomes $(s\{Y'w_1A_iw_2E\})$. Now the only rule that can be applied is of type $P_1(10)$. We apply a rule of type $P_1(10)$ so that A_i is replaced with x, where $A \rightarrow x \in m_i$, and hence the configuration becomes $(s_1\{Y'w_1xw_2E\})$. Finally we apply a rule of type $P_1(12)$ so that Y' is replaced with Y. Now the configuration of the system becomes $(s\{Yw_1xw_2E\})$, which represents the correct simulation of the matrix.

Simulation of a matrix m_i of type 4 is almost same as that of a matrix of type 2. Instead of applying a rule of type $P_1(7)$, we apply a rule of type $P_1(8)$ and erase X_i from the string. Also, after replacing A_i with x by using a rule of type $P_1(11)$, we apply the only applicable rule $P_1(13)$ that erases the symbol E and sends the resulting string w_1xw_2 to the environment. If w_1xw_2 contains any nonterminal, it is not considered as a result.

Since the simulation of a matrix of type 3(1) is similar to that of a matrix of type 3(2), here we explain the process of simulating a matrix m'_i of type 3(1) only. Using a rule of type $P_1(15)$, we first replace X with Y, where $X \rightarrow Y \in m'_i$. Since the cell is now in state $s^{(1)}$, if the symbol $B^{(1)}$ present in the string, it will be replaced with a trap symbol Z using the rule $P_1(16)$. Otherwise, we replace $s^{(1)}$ with s by applying the rule $P_1(17)$ and complete the simulation of the matrix.

The procedure for simulating a matrix of type 2 (type 3) can be iterated for all the matrices of type 2 (type 3). Finally the simulation process ends by simulating a type 4 matrix. So any derivation in G can be simulated by a computation in Π and, conversely, the result of a successful computation in Π corresponds to a terminal derivation in G. Thus, the equality $L(\Pi) = L(G)$ follows. □

As long as universality results are proved using matrix grammars in strong binary normal form, the above result is the optimal one, because to simulate matrices of type $3(j)$, $j \in \{1, 2\}$ we need at least two states and for simulating matrices of other types we need at least another two states.

From the definition of context-free state grammars, one can easily prove that the family of languages generated by single cell rewriting tissue P systems working in *min* mode (for a single cell systems, all three transmission modes are same) is equivalent to the family of languages generated by context-free state grammars. From Theorem 1, it is clear that with leftmost derivation, context-free state grammars with only four states can characterize recursively enumerable languages.

The following theorem shows that systems having $(c, s) \in \{(2, 2), (2, 3), (3, 2)\}$, where c is the number of cells in the system and s is the maximum number of states in any cell of the system, are computationally universal.

Theorem 2. *Let* $n = max\{n_i \mid n_i$ *is the number of states in a cell* σ_i, $1 \leq i \leq m\}$. *Then for all* $m, n < 4$ *such that* $\Sigma_{i=1}^{m} n_i = 4$, $RE = RtP_{m,n}(min, one, left)$.

Proof: Proof follows similar lines. For complete proof one can refer to [5]. □

We now show that systems with four cells and one state in each cell are computationally universal.

Theorem 3. $RE = RtP_{4,1}(min, one, left)$.

Proof: We construct the tissue P system with leftmost derivation

$$\Pi = (V, T, \sigma_1, \sigma_2, \sigma_3, \sigma_4, \{(1,2), (2,3), (3,4), (4,1)\}, 1),$$

where

- $V = T \cup N_1 \cup N_2 \cup \{X_i \mid X \in N_1, 1 \leq i \leq k+1\} \cup \{A_i \mid A \in N_2, 1 \leq i \leq k+1\} \cup \{X^{(1)}, X^{(2)}, X', X'', X''', X'''' \mid X \in N_1\} \cup \{A' \mid A \in N_2\} \cup \{f', f'', f''', f'''', g, h, E, Z\}$;
- $\sigma_1 = (\{s\}, s, \{XAE\}, P_1)$, where P_1 contains the following rules:
  1. $sX \rightarrow sX_1(go), X \in N_1$;
  2. $sX_i \rightarrow sX_{i+1}(go), X \in N_1, 1 \leq i \leq (k-1)$;
  3. $sX_{k+1} \rightarrow sZ(out), X \in N_1$;
  4. $sX_i \rightarrow sY'(go)$, for type 2 matrix $m_i : (X \rightarrow Y, A \rightarrow x)$;
  5. $sX_i \rightarrow sf'(go)$, for type 4 matrix $m_i : (X \rightarrow \lambda, A \rightarrow x)$;
  6. $sA' \rightarrow sA, A \in N_2$;
  7. $sg \rightarrow s\lambda(go)$;
  8. $sX \rightarrow sY^{(j)}(go)$, for $m'_i : (X \rightarrow Y, B^{(j)} \rightarrow \dagger), j \in \{1, 2\}$;
  9. $sE \rightarrow s\lambda(out)$;
- $\sigma_2 = (\{s_1\}, s_1, \emptyset, P_2)$, where P_2 contains the following rules:
  1. $s_1Y'' \rightarrow s_1Y'''(go), Y \in N_1 \cup \{f\}$;
  2. $s_1A \rightarrow s_1A', A \in N_2$;
  3. $s_1A \rightarrow s_1A_1(go), A \in N_2$;
  4. $s_1A_i \rightarrow s_1A_{i+1}(go), A \in N_2, 1 \leq i \leq (k-1)$;
  5. $s_1A_{k+1} \rightarrow sZ(out), A \in N_2$;
  6. $s_1A_i \rightarrow s_1gx(go)$, for type 2 (or type 4) matrix $m_i : (X \rightarrow Y, A \rightarrow x), Y \in N_1 \cup \{\lambda\}$;
  7. $s_1g \rightarrow s_1Z(go)$;
  8. $s_1E \rightarrow s_1Z(go)$;
  9. $s_1Y^{(j)} \rightarrow s_1Y^{(j)}(go), j \in \{1, 2\}$;
- $\sigma_3 = (\{s^{(1)}\}, s^{(1)}, \emptyset, P_3)$, where P_3 contains the following rules:
  1. $s^{(1)}Y''' \rightarrow s^{(1)}Y''''(go), Y \in N_1 \cup \{f\}$;
  2. $s^{(1)}X_i \rightarrow s^{(1)}X_i(go), X \in N_1, 1 \leq i \leq k$;
  3. $s^{(1)}A_i \rightarrow s^{(1)}Z(go), A \in N_2$;
  4. $s^{(1)}g \rightarrow s^{(1)}g(go)$;
  5. $s^{(1)}Y^{(1)} \rightarrow s^{(1)}Yh$;
  6. $s^{(1)}Y^{(2)} \rightarrow s^{(1)}Y^{(2)}(go)$;

7. $s^{(1)}B^{(1)} \rightarrow s^{(1)}Z$;
8. $s^{(1)}E \rightarrow s^{(1)}E(go)$;

- $\sigma_4 = (\{s^{(2)}\}, s^{(2)}, \emptyset, P_4)$, where P_4 contains the following rules:

1. $s^{(2)}Y'''' \rightarrow s^{(2)}Y(go)$, $Y \in N_1$;
2. $s^{(2)}f'''' \rightarrow s^{(2)}\lambda(go)$;
3. $s^{(2)}X_i \rightarrow s^{(2)}X_i(go)$, $X \in N_1$, $1 \leq i \leq k$;
4. $s^{(2)}Y' \rightarrow s^{(2)}Y''(go)$, $Y \in N_1 \cup \{f\}$;
5. $s^{(2)}Z \rightarrow s^{(2)}Z(go)$;
6. $s^{(2)}h \rightarrow s^{(2)}\lambda(go)$;
7. $s^{(2)}Y^{(2)} \rightarrow s^{(2)}Y$;
8. $s^{(2)}B^{(2)} \rightarrow s^{(2)}Z$;
9. $s^{(2)}E \rightarrow s^{(2)}E(go)$.

The system works as follows:

Initially, cell 1 is in the state s and contains XAE. Assume that at a particular instant cell 1 has Xw_1Aw_2E. To simulate a type 2 matrix $m_i : (X \rightarrow Y, A \rightarrow x)$, we first apply a rule of type $P_1(1)$ and send the resulting string $X_1w_1Aw_2E$ to cell 2. In cell 2, as explained in the proof of Theorem 1, we apply rules of type $P_2(2)$ on some of the leftmost nonterminals from the set N_2. By applying a rule of type $P_2(3)$ we send the string $X_1w'_1A_1w_2E$ to cell 3, where w' represents a string in which every nonterminal B is replaced with B'. Now the subscripts of both X and A are incremented by moving the string from cell 1 to cell 2 and from cell 2 to cell 1 (via cell 3 and cell 4), respectively. Assume that, after certain number rewritings, the string becomes $X_iw'_1A_iw_2E$. At any movement we can replace X_i with Y' such that $X \rightarrow Y \in m_i$ (in cell 1) or A_i with gx such that $A \rightarrow x \in m_i$ (in cell 2). If we replace A_i with gx first, the string $X_iw'_1gxw_2E$ will come back to cell 2 (via cell 3, cell 4, and cell 1) as either $X_{i+1}w'_1gxw_2E$ or $Y'w'_1gxw_2E$. In any case, due to the presence of g, a trap symbol will be introduced in cell 2. On the other hand, if we replace X_i with Y' (in cell 1) without replacing A_i with gx (in cell 2), a trap symbol will be introduced in cell 3 due to the absence of X_i and the presence of A_{i+1}. So, for the correct simulation, we first replace X_i with Y' in cell 1 (thus, the string $Y'w'_1A_iw_2E$ is sent to cell 2) and immediately replace A_i with gx in cell 2. After replacing A_i with gx, the resulting string, i.e., $Y'w'_1gxw_2E$, is sent to cell 4 via cell 3. In cell 4, we replace Y' with Y'' and send the string to cell 1. In cell 1, due to the absence of X or X_i, $1 \leq i \leq k+1$, all A's, $A \in N_2$, are replaced with As. Finally, we apply the rule $P_1(7)$ and send the string $Y''w_1xw_2E$ to cell 2. After applying a sequence of rules, i.e., $P_2(1)$, $P_3(1)$, and $P_4(1)$, we get the resulting string Yw_1xw_2E in cell 1, which completes the correct simulation of the matrix m_i on the string Xw_1Aw_2E.

Simulation of a matrix m_i of type 4 is similar to that of a matrix of type 2 with few changes in the procedure. Instead of replacing X_i with Y', we replace X_i with f' and send the string to cell 2. In cell 2, we replace A_i with gx and send the string to cell 4 via cell 3. Application of a rule of type $P_4(4)$ replaces f' with f'' sends the resulting string to cell 1. After applying a sequence of rules, i.e., $P_1(7)$, $P_2(1)$, $P_3(1)$, and $P_4(1)$, we get the resulting string w_1xw_2E in cell 1. Finally we apply the rule $P_1(9)$ and send the string w_1xw_2 to the environment, which completes the correct simulation of the matrix of type 4.

Simulation of a matrix of type $3(j)$, $j \in \{1, 2\}$ is as follows. Since cell 3 has the rules for simulating matrices of type 3(1) and cell 4 has the rules for simulating matrices of type 3(2), whenever X is replaced with $Y^{(1)}$ in cell 1, the resulting string is sent to cell 3 via cell 2. In cell 3, we replace $Y^{(1)}$ with Y and check for $B^{(1)}$. If $B^{(1)}$ is presented in the string, a trap symbol will be introduced. Otherwise, we send the string to cell 1 via cell 4. If cell 2 gets a string of the form $Y^{(2)}wE$ from cell 1, we send the string to cell 4 via cell 3. In cell 4, we replace $Y^{(2)}$ with Y and check for $B^{(2)}$. If $B^{(2)}$ is presented in the string, a trap symbol will be introduced. Otherwise, we send the string to cell 1.

We repeat the process of simulating matrices of type 2 and type 3 for all such type of matrices of G and complete the process by simulating a type 4 matrix. So any derivation in G can be simulated by a computation in Π and, conversely, the result of a successful computation in Π corresponds to a terminal derivation in G. Thus, the equality $L(\Pi) = L(G)$ follows. □

Like Theorem 1, this result is also the optimal one in the case where universality results are proved using matrix grammars in strong binary normal form. The reason is as follows: we need at least two cells for simulating matrices of type $3(j)$, $j \in \{1, 2\}$ and another two cells for simulating matrices of other types.

4 Conclusion

We considered tissue P systems with leftmost derivation and gave universality results. As one-cell tissue P systems are equivalent to context-free state grammars, it is clear that with leftmost rewriting restriction, context-free state grammars with four states are computationally universal. In all the proofs, the total number of states used in the tissue P system is 4. It is not clear whether or not this *magic number* can be further reduced. From the definition of the tissue P system, we understand that each cell has at least one state. So, for any system where the total number of states used is 4, the minimum and the maximum number of cells must be 1 and 4, respectively. Four states can be distributed to i, $1 \leq i \leq 4$, cells in following ways: if the number of cells is 1, that cell has all four states; if the number of cells is 2, four states can be distributed in two ways: 1) one cell has one state and the other cell has three states, or 2) both cells have two states; if the number of cells is 3, four states can be distributed in only one way, i.e., one cell has two states and the other two cells have one state each; and if the number is 4, each cell has only one state. From Section 3 it is clear that the universality results are given for all possible distributions of four states to all available cells. From these universality results, we conjecture that if tissue P systems with leftmost derivation with k states, where $k \leq 3$, are computationally universal, then any tissue P system with leftmost derivation having (m, n) cells and states pair is computationally universal for all possible values of m and n such that $1 \leq m, n \leq k$ and $\Sigma_{i=1}^{m} n_i = 3$, where n_i is the number of states in a cell σ_i and $n = max\{n_i \mid 1 \leq i \leq m\}$. It is an interesting research problem to check whether or not a tissue P system with leftmost derivation Π_1 can be simulated by another tissue P system with leftmost derivation Π_2 such that the total number of states in Π_1 is equal to that of Π_2.

References

[1] `http://psystems.disco.unimib.it/`

[2] J. Dassow, Gh. Păun: Regulated Rewriting in Formal Language Theory. Springer, Berlin, 1989.

[3] R. Freund, Gh. Păun: On the number of non-terminal symbols in graph-controlled, programmed, and matrix grammars. in: M. Margenstern and Y. Rogozhin (Eds.), *Machines, Computations, and Universality*, 3rd Internat. Conf. Lecture Notes in Computer Science, **2055**, Springer, Berlin, 2001, 214–225.

[4] M. Mutyam, V.J. Prakash, K. Krithivasan: Rewriting tissue P systems. *J. Univ. Comput. Sci.* **10(9)** (2004) 1250–1271.

[5] M. Mutyam, K. Krithivasan: Tissue P systems with leftmost derivation. IIIT Technical Report, 2005.

[6] C. Martín-Vide, Gh. Păun, J. Pazos, A. Rodríguez-Patón: Tissue P systems. *Theoret. Comput. Sci.* **296(2)** (2003) 295–326.

[7] Gh. Păun: Computing with membranes. *J. Comput. System Sci.* **61(1)** (2000) 108–143.

[8] Gh. Păun: Membrane Computing. An Introduction, Springer, Berlin, 2002.

[9] V.J. Prakash, K. Krithivasan: Leftmost rewriting tissue P systems. in: 6th Internat. Workshop on Descriptional Complexity of Formal Systems, London, Ontario, Canada, 2004.

Madhu Mutyam
International Institute of Information Technology,
Hyderabad - 19, India
email: `madhu_mutyam@iiit.ac.in`

Kamala Krithivasan
Department of Computer Science and Engineering,
Indian Institute of Technology Madras,
Chennai - 36, India
email: `kamala@iitm.ernet.in`

Formal Language Aspects of Natural Computing– 2006, pp. 197–207

Hexagonal Array Splicing Systems

D.G. Thomas, M.H. Begam and N.G. David

1 Introduction

In studies of picture generation, several grammars were introduced in the literature to generate various classes of pictures. One such grammar is the hexagonal kolam array grammar (HKAG) by Siromoneys [6] generating a class of hexagonal pictures (arrays) (HKAL). Based on the study [6], Subramanian [7] considered a hierarchy of hexagonal array grammars (HAG) generating classes of hexagonal arrays on triangular grids (HAL). Recently Dersanambika et al [1] have defined local and recognizable hexagonal picture languages (HLOC and HREC) using hexagonal tiling system and studied the properties of these classes, thus providing an analog of the corresponding rectangular picture languages studied in [2].

On the other hand, splicing systems were recently introduced by Head [3] on biological considerations to model certain recombinant behaviour of DNA molecules and are of current interest and study [4]. The splicing systems make use of a new operation called splicing on strings of symbols. This splicing operation has been recently applied to rectangular arrays [5] and graphs [2].

In this paper, we introduce a new method of applying the splicing operation on images of hexagonal arrays. Splicing rules that involve 2×1 or 1×2 dominoes along x or y directions are considered. Two hexagonal arrays are x-direction spliced or y-direction spliced by using the domino splicing rules in parallel. The resulting model called hexagonal array splicing system (HASS) is simple to handle and generate a new class of hexagonal array languages (HASSL). We compare the class HASSL with other classes HLOC, HREC, HKAL and bring out their differences. Some closure properties of HASSL under language theoretic operation union and geometric operation reflections about the base lines are considered. This study initiated in this paper might prove useful to analyze better the structure of images.

2 Preliminaries

We consider hexagons of the type :

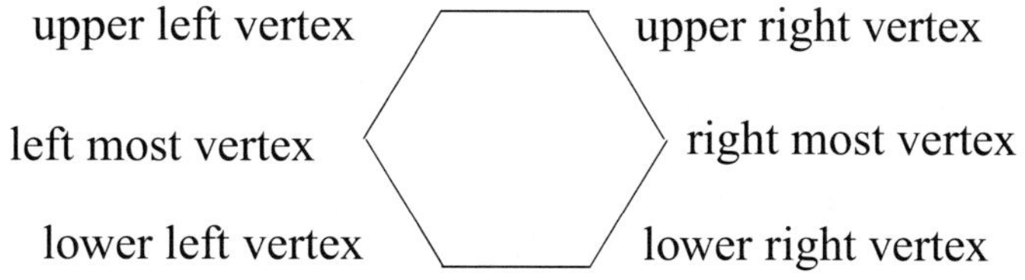

Let Σ be a finite alphabet of symbols. A hexagonal picture p over Σ is a hexagonal array of symbols of Σ. For example, a hexagonal picture over the alphabet $\{a, b\}$ is :

$$\begin{matrix} & a & & a & \\ a & & b & & b \\ & b & & a & \end{matrix}$$

The set of all hexagonal arrays of the alphabet Σ is denoted by Σ^{**H}. A hexagonal picture language L over Σ is a subset of Σ^{**H}.

With respect to a triad 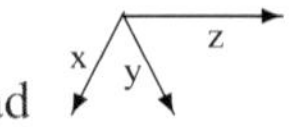 of triangular axes x, y, z the coordinates of each element of hexagonal picture can be fixed.

For $p \in \Sigma^{**H}$, let $\hat{p}$ be the hexagonal array obtained by surrounding p with a special boundary symbol $\# \notin \Sigma$.

$$
\begin{array}{ccccccccc}
 & & \# & & \# & & \# & & \\
 & \# & & a & & a & & \# & \\
\hat{p} = \# & & a & & b & & b & & \# \\
 & \# & & b & & a & & \# & \\
 & & \# & & \# & & \# & &
\end{array}
$$

Fig 1.

Given a picture $p \in \Sigma^{**H}$, let $l_1(p)$ denote the number of elements in the border of p from upper left vertex to left most vertex in the direction $\swarrow$ called x direction, $l_2(p)$ denote the number of elements in the border of p from upper right vertex to right most vertex in the direction $\searrow$ called y direction and $l_3(p)$ denote the number of elements in the border of p from upper left vertex to upper right vertex in the direction $\rightarrow$ called z direction.

The directions are fixed with origin of reference as the upper left vertex, having coordinates $(1, 1, 1)$. The triple $(l_1(p), l_2(p), l_3(p))$ is called the size of the picture p. Furthermore, if $1 \leq i \leq l_1(p)$, $1 \leq j \leq l_2(p)$, $1 \leq k \leq l_3(p)$, let p_{ijk} denote the symbol in p with coordinates (i, j, k). For example, the hexagonal array given in Fig. 1 is of size $(2, 2, 2)$ and $p_{111} = a$, $p_{221} = b$, etc.

Given a hexagonal picture p of size (l, m, n) for $g \leq l$, $h \leq m$ and $k \leq n$, we denote by $B_{g,h,k}(p)$ the set of all hexagonal blocks (or hexagonal subpictures) of p of size (g, h, k). Each member of $B_{2,2,2}(p)$ is called a hexagonal tile.

Recently, Dersanambika et al have introduced two interesting classes of hexagonal picture languages, viz., the class of local hexagonal picture languages (HLOC) and the class of recognizable hexagonal picture languages (HREC).

The family HLOC

Let Σ be a finite alphabet. A hexagonal picture language $L \subseteq \Sigma^{**H}$ is called local if there exists a finite set Δ of hexagonal tiles over $\Sigma \cup \{\#\}$ such that $L = \{p \in \Sigma^{**H} / B_{2,2,2}(\hat{p}) \subseteq \Delta\}$.

The family of hexagonal local picture languages will be denoted by HLOC.

The family HREC

Let Σ be a finite alphabet. A hexagonal picture language $L \subseteq \Sigma^{**H}$ is called recognizable if there exists a hexagonal local picture language L' over an alphabet Γ and a mapping $\pi : \Gamma \rightarrow \Sigma$ such that $L = \pi(L')$.

Example 1. *Let* $\Sigma = \{1, 2, 3\}$;
Then

$$
L_1 = \left\{ \begin{array}{c} 1\ 1 \\ 2\ 2\ 2 \\ 3\ 3 \end{array},\ \begin{array}{c} 1\ 1\ 1 \\ 2\ 2\ 2\ 2 \\ 3\ 3\ 3 \end{array},\ \begin{array}{c} 1\ 1\ 1\ 1 \\ 2\ 2\ 2\ 2\ 2 \\ 3\ 3\ 3\ 3 \end{array}, \ldots \right\} \quad \textit{is local}
$$

Example 2. *Let $\Sigma = \{a\}$. It is shown [1] that the language of hexagonal pictures over Σ with all sides of equal length is not local, but, recognizable.*

3 Hexagonal Array Splicing Systems

We introduce hexagonal array splicing systems.

Let Σ be an alphabet. Let ₵, \$ be two special symbols, not in Σ. A domino over Σ is of the form

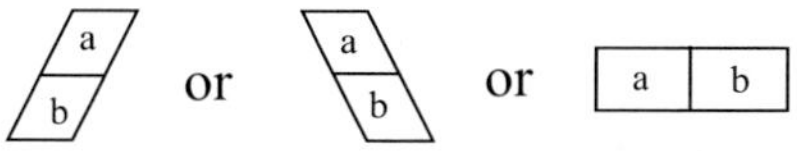

, a,b ε Σ

A domino x direction splicing rule over Σ is of the form $r = \alpha_1 ₵ \alpha_2 \$ \alpha_3 ₵ \alpha_4$ where each

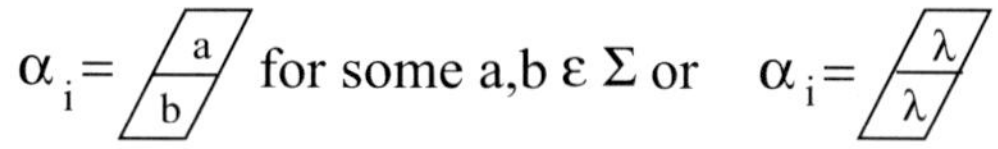

where λ is the empty word ($1 \le i \le 4$).

A domino y direction splicing rule over Σ is of the form $r = \beta_1 ₵ \beta_2 \$ \beta_3 ₵ \beta_4$ where each

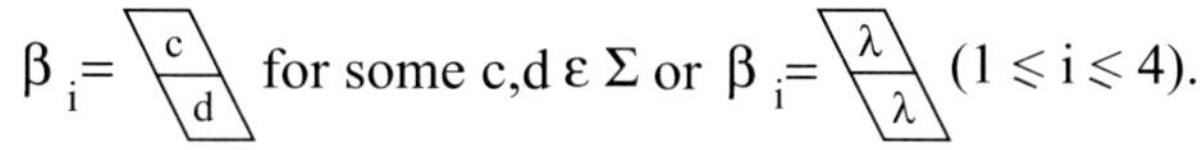

Given two hexagonal arrays X and Y of sizes (l, m, n) and (l, m, n') respectively, i.e., $X =< a_{ijk} >$, $1 \le i \le l$, $1 \le j \le m$, $1 \le k \le n$; $Y =< b_{i'j'k'} >$, $1 \le i' \le \ell$, $1 \le j' \le m$, $1 \le k' \le n'$, we write $(X, Y) \overset{x}{\vdash} W$ if there exist x direction splicing rules $r_1, r_2, \ldots, r_p$ not all different such that each r_i is of the form

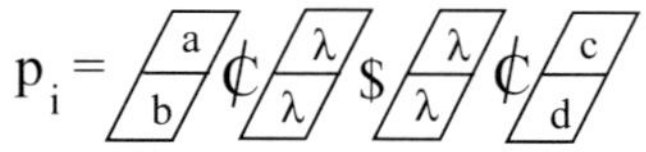

and $z =< c_{ijk'} >, 1 \le i \le l, 1 \le j \le m,$

$1 \le k' \le n'', n'' \ge n'$ and n

We now say that W is obtained from X and Y by domino x direction splicing in parallel.

We can similarly define domino y direction splicing in parallel.

We first define a Z-hexagonal array scheme and a Z-hexagonal array splicing system. This Z-hexagonal array splicing system generates the language of hexagonal arrays p of sizes $(\ell_1(p), \ell_2(p), \ell_3(p))$ where $\ell_1(p)$ and $\ell_2(p)$ are fixed and $\ell_3(p)$ varies.

A Z-hexagonal array scheme is a triplet $\Gamma_Z = (\Sigma, R_x^Z, R_y^Z)$ where Σ is an alphabet, R_x^Z is a finite set of domino x direction splicing rules and R_y^Z, a finite set of domino y direction splicing rules. For a given hexagonal array scheme $\Gamma_Z = (\Sigma, R_x^Z, R_y^Z)$ and a language $L \subseteq \Sigma^{**H}$, we define

$$\Gamma_Z(L) = \{W \in \Sigma^{**H} / (X, Y) \overset{x}{\vdash} W \text{ or } (X, Y) \overset{y}{\vdash} W, \text{ for some } X, Y \in L,$$

$p_i \in R_x^Z$ and $q_j \in R_y^Z, 1 \leq i \leq m, 1 \leq j \leq n\}$

Iteratively, we define $\Gamma_Z^{\circ}(L) = L$, $\Gamma_Z^{i+1}(L) = \Gamma_Z(L) \cup \Gamma_Z(\Gamma_Z(L))$, $i \geq 0$, $\Gamma_Z^*(L) = \bigcup_{i \geq 0} \Gamma_Z^i(L)$. A Z-hexagonal array splicing system is defined by $S = (\Gamma_Z, I)$ where $\Gamma_Z = (\Sigma, R_x^Z, R_y^Z)$ and I is a finite subset of Σ^{**H}. The language of S is defined by $L(S) = \Gamma_Z^*(I)$ and we call it a Z-hexagonal array splicing language and denote the class of such languages by Z-HASSL.

Example 3. *Let* $\Sigma = \{a\}$, $I = \left\{ \begin{array}{c} a\ a \\ a\ a\ a \\ a\ a \end{array} \right\}$

Let $S = (\Gamma_Z, I)$ *where* $\Gamma_Z = \left(\Sigma, R_x^Z, R_y^Z\right)$ *with*

$R_x^Z = \{ p_1 :$ 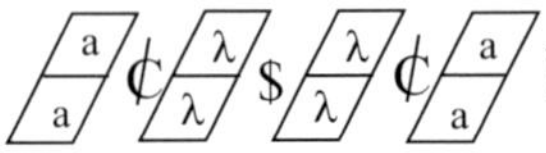 $\}$

$R_y^Z = \phi.$

Using x direction splicing rule p_1 *in parallel,*

$$\left(\begin{array}{c} a\ a/\lambda \\ a\ a/\lambda\ a \\ a/\lambda\ a \end{array} , \begin{array}{c} a\ \lambda/a \\ a\ \lambda/a\ a \\ \lambda/a\ a \end{array} \right) \vdash_x \begin{array}{c} a\ a\ a \\ a\ a\ a\ a \\ a\ a\ a \end{array}$$

$$L(S) = \left\{ \begin{array}{c} a\ a \\ a\ a\ a \\ a\ a \end{array} , \begin{array}{c} a\ a\ a \\ a\ a\ a\ a \\ a\ a\ a \end{array} , \ldots \right\}$$

Remark 1. We can similarly define X-hexagonal array splicing scheme and a X-hexagonal array splicing system.

A X-hexagonal array scheme is a triplet $\Gamma_X = (\Sigma, R_x^X, R_y^X)$ where Σ is an alphabet, R_x^X is a finite set of domino x direction splicing rules and R_y^X, a finite set of domino y direction splicing rules.

A X-hexagonal array splicing system is defined by $S = (\Gamma_X, I)$ where $\Gamma_X = (\Sigma, R_x^X, R_y^X)$ and I is a finite subset of Σ^{**H}. Here the splicing is done as follows:

Given two hexagonal arrays X and Y of sizes (ℓ, m, n) and (ℓ', m, n)

(i) we first rotate X and Y, $60°$ in the clockwise direction about the Z-axis passing through the upper left vertex, keeping the upper left vertex fixed and get two new hexagonal arrays X' and Y'. For example if $X = \begin{array}{c} a\ b \\ c\ d\ e \\ f\ g \end{array}$, then X' is $\begin{array}{c} c\ a \\ f\ d\ b \\ g\ e \end{array}$.

(ii) Then we obtain a hexagonal array W' from X' and Y' by either x direction splicing in parallel using x direction splicing rules in R_x^X or y direction splicing in parallel using y direction splicing rules in R_y^X

(iii) Finally, by rotating W' $60°$ in the anticlockwise direction about the Z-axis passing through the upper left vertex keeping the upper left vertex fixed, we get a new hexagonal array W. We say that W is obtained from X and Y using S. We thus define $\Gamma_X(L)$

and $\Gamma_X^*(L)$ similar to $\Gamma_Z(L)$ and $\Gamma_Z^*(L)$. This X-hexagonal array splicing system generates the language of hexagonal arrays p of sizes $(\ell_1(p), \ell_2(p), \ell_3(p))$ where $\ell_2(p)$ and $\ell_3(p)$ are fixed and $\ell_1(p)$ varies.

Remark 2. Similarly, we can define Y-hexagonal array splicing scheme and a Y-hexagonal array splicing system.

A Y-hexagonal array scheme is a triplet $\Gamma_Y = (\Sigma, R_x^Y, R_y^Y)$ where Σ is an alphabet, R_x^Y is a finite set of domino x direction splicing rules and R_y^Y, a finite set of domino y direction splicing rules.

A Y-hexagonal array splicing system is defined by $S = (\Gamma_Y, I)$ where $\Gamma_Y = (\Sigma, R_x^Y, R_y^Y)$ and I is a finite subset of Σ^{**H}. Here the splicing is done as follows:

Given two hexagonal arrays X and Y of sizes (ℓ, m, n) and (ℓ, m', n)

(i) we first rotate X and Y, 60° in the anticlockwise direction about the Z-axis passing through the upper left vertex, keeping the upper left vertex fixed and get two new hexagonal arrays X' and Y'. For example if $X = \begin{matrix} a\ b \\ c\ d\ e \\ f\ g \end{matrix}$, then X' is $\begin{matrix} b\ e \\ a\ d\ g \\ c\ f \end{matrix}$

(ii) Then we obtain a hexagonal array W' from X' and Y' by either x direction splicing in parallel using x direction splicing rules in R_x^Y or y direction splicing in parallel using y direction splicing rules in R_y^Y

(iii) Finally, by rotating W', 60° in the clockwise direction about the Z-axis passing through the upper left vertex keeping the upper left vertex fixed, we get a new hexagonal array W. We say that W is obtained from X and Y using S. We thus define $\Gamma_Y(L)$ and $\Gamma_Y^*(L)$ similar to $\Gamma_Z(L)$ and $\Gamma_Z^*(L)$. This Y-hexagonal array splicing system generates the language of hexagonal arrays p of sizes $(\ell_1(p), \ell_2(p), \ell_3(p))$ where $\ell_1(p)$ and $\ell_3(p)$ are fixed and $\ell_2(p)$ varies.

Definition 1. A hexagonal array splicing scheme is a triplet $\Gamma = (\Gamma_X, \Gamma_Y, \Gamma_Z)$ with alphabets of Γ_X, Γ_Y and Γ_Z being the same Σ. We can define $\Gamma^0(L) = L, \Gamma(L) = \Gamma_X(L) \cup \Gamma_Y(L) \cup \Gamma_Z(L)$.

$$\Gamma^*(L) = \Gamma^0(L) \cup \left(\bigcup_{\substack{i \geq 1 \\ D, D' \in \{X, Y, Z\}}} \Gamma_D(\Gamma_{D'}^{i-1}(L)) \right)$$

A hexagonal array splicing system is $S = (\Gamma, I)$ and the hexagonal array splicing language is $L(S) = \Gamma^*(I)$, where I is the set of axioms. We denote the class of such hexagonal array splicing languages by HASSL.

In the following theorems, we deal with hexagonal array splicing systems $S = (\Gamma, I)$ with $\Gamma_X = (\Sigma, \phi, \phi)$ and $\Gamma_Y = (\Sigma, \phi, \phi)$ where ϕ is the empty set.

Theorem 1. *The classes of HLOC and HASSL are incomparable but not disjoint.*

Proof: Consider the hexagonal local language L given in [1]. A member of L is the hexagonal array

$$\begin{array}{c} 1\ 0\ 0 \\ 0\ 0\ 1\ 0 \\ 0\ 1\ 0\ 0\ 1 \\ 0\ 0\ 1\ 0 \\ 1\ 0\ 0 \end{array}$$

Even though $L \in HLOC$, $L \notin HASSL$, since it is clear from the definition of x direction and y direction splicing that only hexagonal arrays of size (l, m, n) with $l = m = n$ cannot be generated.

The language given in example 3 is in HASSL, but is not in HLOC.

The language given in example 1 is in HLOC. It is also in HASSL, for, if, $\Sigma = \{1, 2, 3\}$,

$$I = \left\{ \begin{array}{c} 1\ 1 \\ 2\ 2\ 2 \\ 3\ 3 \end{array} \right\}, S = (\Gamma, I) \text{ where } \Gamma_Z = (\Sigma, R_x^Z, R_y^Z) \text{ with}$$

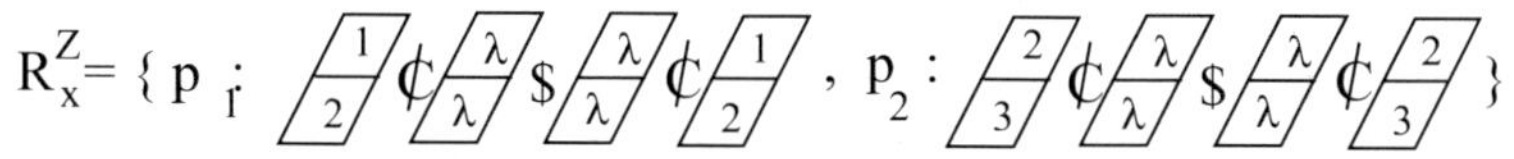

$$R_y^Z = \phi,$$

we get, $L(S) = L_1$.

For example

$$\left(\begin{array}{c} 1\ 1/\lambda \\ 2\ 2/\lambda\ 2 \\ 3/\lambda\ 3 \end{array}, \begin{array}{c} 1\ \lambda/1 \\ 2\ \lambda/2\ 2 \\ \lambda/3\ 3 \end{array} \right) \overset{x}{\vdash} \left(\begin{array}{c} 1\ 1\ 1 \\ 2\ 2\ 2\ 2 \\ 3\ 3\ 3 \end{array} \right)$$

□

Theorem 2. *The class HASSL is not closed under union.*

Proof: Let L_4 be a language of hexagonal arrays of sizes (l, m, n) with $l = m = 2$ and $n \geq 4$, west (left) triangular border made up of a's and east (right) triangular border made up of b's and inner part made up of v's. A member of L_4 is shown below:

$$\begin{array}{c} a\ v\ v\ b \\ a\ v\ v\ v\ b \\ a\ v\ v\ b \end{array}$$

Similarly L_5 is another language of hexagonal arrays as in L_4 but west and east triangular borders are made up of c's and d's respectively with inner parts made up of v's. It is clear that $L_4, L_5 \in Z - HASSL$. A member of $L_4 \cup L_5$ will have west and east triangular borders made up of only a's and b's respectively or only c's and d's respectively. But, we cannot find any new hexagonal array splicing system to generate $L_4 \cup L_5$. □

Remark 3.

1. Given a hexagonal array α, we can obtain a new hexagonal array denoted by $rh(\alpha)$, by taking reflection of α about a horizontal line.

2. Similarly, given a hexagonal array α, we can obtain a new hexagonal array, denoted by $rv(\alpha)$, by taking reflection of α about a vertical line.

$$\text{For example if } \alpha = \begin{array}{c} a\ b\ c \\ d\ e\ f\ g \\ h\ i\ j\ k\ l \\ m\ n\ o\ p \\ q\ r\ s \end{array} \quad \text{then} \quad rh(\alpha) = \begin{array}{c} q\ r\ s \\ m\ n\ o\ p \\ h\ i\ j\ k\ l \\ d\ e\ f\ g \\ a\ b\ c \end{array}$$

$$\text{and } rv(\alpha) = \begin{array}{c} c\ b\ a \\ g\ f\ e\ d \\ l\ k\ j\ i\ h \\ p\ o\ n\ m \\ s\ r\ q \end{array}.$$ If $L \subseteq \Sigma^{**H}$, then, let

$RH(L) = \{rh(\alpha)/\alpha \in L\}$; $RV(L) = \{rv(\alpha)/\alpha \in L\}$

We introduce two more notations:

If $p_1 \in R^Z_x$, let $p_1 =$ 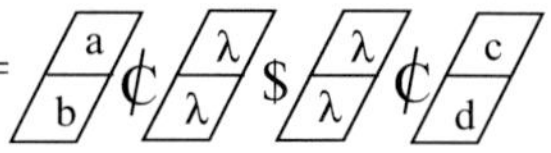

$HDy(p_1) =$ 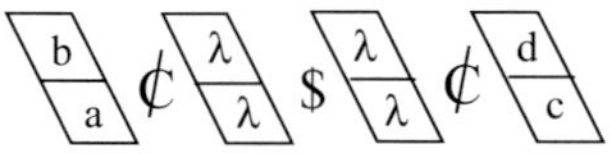

$HDy(R^Z_x) = \{HDy(p) / p \in R^Z_x\}$

Similarly we can define $HDx(R^Z_y)$

If $p_2 \in R^Z_y$, let $p_2 =$ 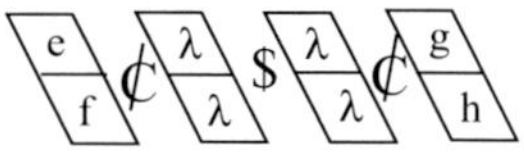

Define $Dx(p_2) =$ 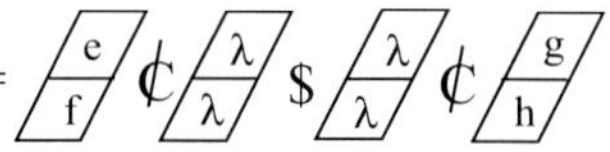

$Dx(R^Z_y) = \{Dx(p) / p \in R^Z_y\}$

Similarly we can define $Dy(R^Z_x)$.

Theorem 3. *Let $S = (\Gamma, I)$ where $\Gamma_Z = (\Sigma, R^Z_x, R^Z_y)$ be a hexagonal array splicing system generating L. Then, RH(L) and RV(L) are generated by hexagonal array splicing systems.*

Proof: $RH(L)$ is generated by (Γ_1, I_1) where $\Gamma_{1_Z} = (\Sigma, HDx(R^Z_y), HDy(R^Z_x))$; $I_1 = RH(I)$
$RV(L)$ is generated by (Γ_2, I_2) where $\Gamma_{2_Z} = (\Sigma, Dy(R^Z_x), Dx(R^Z_y))$; $I_2 = RV(I)$ □

4 Other Comparison Results

Siromoney et al [6] introduced two classes of hexagonal array languages, viz., hexagonal kolam array grammar languages (HKAGL) and controlled table OL hexagonal array languages (CTOLHAL). For definitions and notations, we refer to [6]. In this section, we compare these classes with the HASSL.

Theorem 4. *The classes of HKAGL and HASSL are incomparable but not disjoint.*

Proof: Let $\mathcal{G}_1 = (V, I, P, S, \mathcal{L})$ be a hexagonal kolam array grammar where $V = V_1 \cup V_2$ with $V_1 = \{S, S_1\}$, $V_2 = \{a, b, c, x, y, z\}$, $I = \{G, Y\}$ where G stands for grey and Y for yellow, $P = P_1 \cup P_2$ with

$$P_1 = \{S \rightarrow (((S_1 \oslash a) \obslash b) \leftarrow\!\ominus\, c),\;\; S_1 \rightarrow (((S \oslash x) \obslash y) \leftarrow\!\ominus\, z)\}$$

$$P_2 = \left\{ S \rightarrow \begin{array}{c} G\ G\ G \\ G\ G\ G\ G \\ G\ G\ G\ G\ G \\ G\ G\ G\ G \\ G\ G\ G \end{array} \right\}, \quad \mathcal{L} = \{L_a, L_b, L_c, L_x, L_y, L_z\}$$

with
$L_a = \{G^n <G> G^n\}$, $L_b = \{G^{n+1} <G> G^n\}$
$L_c = \{G^{n+1} <G> G^{n+1}\}$, $L_x = \{Y^n <Y> Y^n\}$
$L_y = \{Y^{n+1} <Y> Y^n\}$, $L_z = \{Y_{n+1} <Y> Y^{n+1}\}$
$(n \geq 1)$ and S is the start symbol.
$\mathcal{G}_1$ generates the set of hexagonal arrays, the first two members of which are given below:

$$\begin{array}{c} G\ G\ G \\ G\ G\ G\ G \\ G\ G\ G\ G\ G \\ G\ G\ G\ G \\ G\ G\ G \end{array},$$

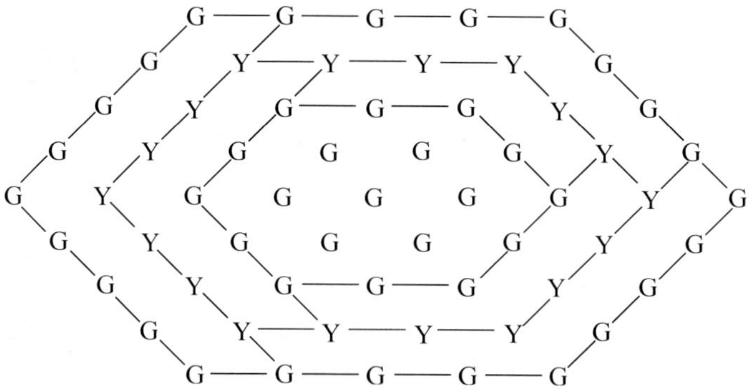

This language cannot be generated by a hexagonal array splicing system.

Consider the following hexagonal kolam array grammar $\mathcal{G}_2 = (V, I, P, L, S)$ where $V = V_1 \cup V_2$, $V_1 = \{S\}$, $V_2 = \{a\}$, $I = \{x\}$, $P = P_1 \cup P_2$ with

$$P_1 = \{S \rightarrow S \ominus a\}, P_2 = \{S \rightarrow \begin{array}{c} x\ x \\ x\ x\ x \\ x\ x \end{array}\}, \mathcal{L} = \{L_a\}$$

where $L_a = \{x^n <x> x^n\}$, $n = 1$
$\mathcal{G}_2$ generates the language,

$$\left\{ \begin{array}{c} x\ x \\ x\ x\ x \\ x\ x \end{array},\; \begin{array}{c} x\ x\ x \\ x\ x\ x\ x \\ x\ x\ x \end{array},\; \begin{array}{c} x\ x\ x\ x \\ x\ x\ x\ x\ x \\ x\ x\ x\ x \end{array}, \ldots \right\}$$

This language can be generated by a hexagonal array splicing system (Γ_1, I_1) with $\Gamma_{1_Z} = (\Sigma, R_x^Z, R_y^Z)$, $\Sigma = \{x\}$, $R_x^Z = \phi$,

$R_y^Z = \{$ 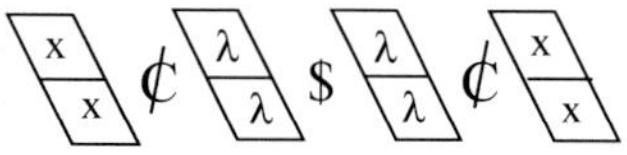 $\}$

and $I_1 = \left\{ \begin{array}{c} x\ x \\ x\ x\ x \\ x\ x \end{array} \right\}$. Hence the classes HKAGL and HASSL are not disjoint.

Consider the hexagonal array splicing system (Γ_2, I_2) where $\Gamma_{2_Z} = (\Sigma, R_x^Z, R_y^Z)$, $\Sigma = \{1, 2, 3\}$,

$R_x^Z = \phi$,

$R_y^Z = \{$ 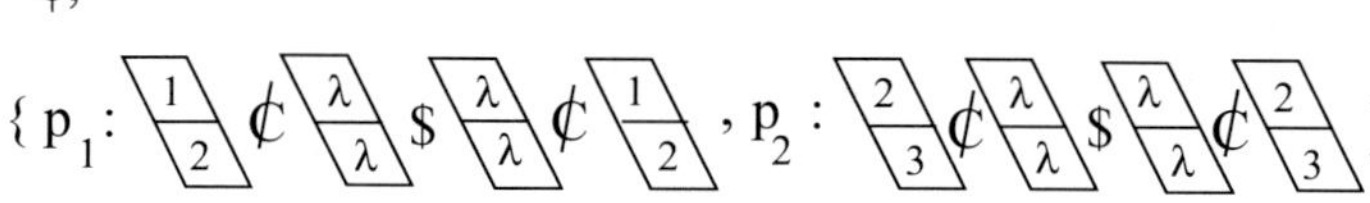 $\}$

$I = \left\{ \begin{array}{c} 1\ 1 \\ 1\ 2\ 1 \\ 3\ 3 \end{array} \right\}$. The language L generated by this splicing system is

$$\left\{ \begin{array}{c} 1\ 1 \\ 1\ 2\ 1 \\ 3\ 3 \end{array}, \begin{array}{c} 1\ 1\ 1 \\ 1\ 2\ 2\ 1 \\ 3\ 3\ 3 \end{array}, \begin{array}{c} 1\ 1\ 1\ 1 \\ 1\ 2\ 2\ 2\ 1 \\ 3\ 3\ 3\ 3 \end{array}, \ldots \right\}$$

We cannot construct a HKAG to generate this language since if we take the rule $S \to \begin{array}{c} 1\ 1 \\ 1\ 2\ 1 \\ 3\ 3 \end{array}$ in P_2, other elements of the language of HKAG are constructed either through the west arrow catenation or east arrow catenation and hence the first element of the language must be either a prefix or a suffix hexagonal array of the second element of the language. In general, the nth element of the language must be a prefix or a suffix array of the $(n+1)^{th}$ element which is not in the language L. □

Theorem 5. *The classes regular CTOLHAL and HASSL are incomparable but not disjoint.*

Proof: Consider the regular controlled table OL hexagonal array grammar $G_1 = (V, \mathcal{P}, H_0, C)$ where $V = \{a\}$, $H_0 = \begin{array}{c} a\ a \\ a\ a\ a \\ a\ a \end{array}$, $\mathcal{P} = \{T_1, T_2, T_3\}$, $T_1 = \{a \nearrow aa\}$, $T_2 = \{a \searrow aa\}$, $T_3 = \{aa \leftarrow a\}$, $C = \{(T_1T_2T_3)^n / n \geq 0\}$. The language generated by this grammar is

$$L_6 = \left\{ \begin{array}{c} a\ a \\ a\ a\ a \\ a\ a \end{array}, \begin{array}{c} a\ a\ a \\ a\ a\ a\ a \\ a\ a\ a\ a\ a \\ a\ a\ a\ a \\ a\ a\ a \end{array}, \ldots \right\}$$

L_6 cannot be generated by any system in HASS.

Let $G_2 = (V, \mathcal{P}, H_0, C)$ where $V = \{a\}$, $H_0 = \begin{array}{ccccc} & a & & a & \\ a & & a & & a \\ & a & & a & \end{array}$, $\mathcal{P} = \{T_1, T_2, T_3\}$, $T_1 = \{a \nearrow a\}$, $T_2 = \{a \searrow a\}$, $T_3 = \{aa \leftarrow a\}$, $C = \{(T_1T_2T_3)^n / n \geq 0\}$.

G_2 generates the language

$$L_7 = \left\{ \begin{array}{ccccc} & a & & a & \\ a & & a & & a \\ & a & & a & \end{array}, \begin{array}{ccccccc} & a & & a & & a & \\ a & & a & & a & & a \\ & a & & a & & a & \end{array}, \ldots \right\}$$

This is generated by a hexagonal array splicing system $S = (\Gamma, I)$ where $\Gamma_Z = (\Sigma, R_x^Z, R_y^Z)$ with $\Sigma = \{1, 2, 3\}$, $I = \begin{array}{ccccc} & 1 & & 1 & \\ 1 & & 2 & & 1 \\ & 3 & & 3 & \end{array}$

$R_x^Z = \phi$,

$R_y^Z = \{ p_1 :$ 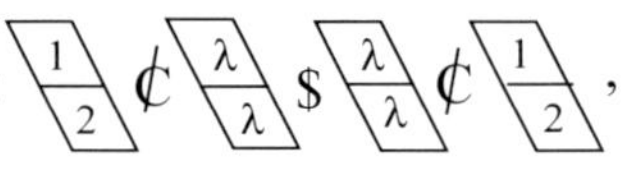 , p_2 : 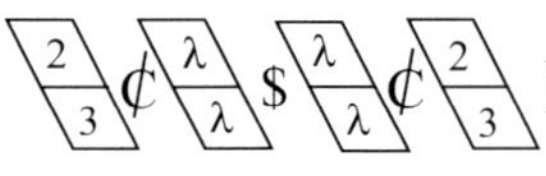 $\}$

One member of this language generated by S is $\begin{array}{ccccccc} & 1 & & 1 & & 1 & \\ 1 & & 2 & & 2 & & 1 \\ & 3 & & 3 & & 3 & \end{array}$ This language cannot be generated by any regular controlled table OL hexagonal array grammar, since, if we allow any of the rule $11 \leftarrow 1$ or $12 \leftarrow 1$, we get a different language. □

5 Conclusion

In this paper we have considered splicing on hexagonal arrays. Some closure properties and results on generative power are discussed. Directions for further exploration are decidability results, using context conditions and rules similar to simple splicing.

References

[1] K.S. Dersanambika, K. Krithivasan, C. Martin-Vide and K.G. Subramanian: Hexagonal Pattern Languages. *Proceedings of International Workshop on Combinatorics Image Analysis*, New Zealand, 2004.

[2] D. Giammarresi and A. Restivo: Two-dimensional Languages. In *Handbook of Formal Languages* **3** Eds. G. Rozenberg and A. Salomaa, Springer-Verlag, 1997.

[3] T. Head: Formal Language Theory and DNA : An analysis of the generative capacity of specific recombinant behaviours. *Bulletin of Mathematical Biology* **49** (1987), 1134–1142.

[4] T. Head, Gh. Paun and D. Pixton: Language Theory and Molecular Genetics : Generative Mechanisms suggested by DNA Recombination. In *Handbook of Formal Languages* **2 Ch. 7**, Eds. G. Rozenberg and A. Salomaa, Springer-Verlag, 1997.

[5] K. Krithivasan, V.T. Chakravarthy, R. Rama: Array splicing systems. In New Trends in Formal Languages Control, Cooperation and Combinatorics (Gh. Paun, A. Salomaa, eds.), *Lecture Notes in Computer Science* **1218** Springer-Verlag, Berlin (1997), 346–365.

[6] G. Siromoney and R. Siromoney: Hexagonal Arrays and Rectangular Blocks. *Computer Graphics and Image Processing* **5** 1976, 353–381.

[7] K.G. Subramanian: Hexagonal Array Grammars. *Computer Graphics and Image Processing* **10(4)**, 1979, 388–394.

D.G. Thomas and N.G. David
Department of Mathematics,
Madras Christian College,
Chennai - 600 059

M.H. Begam
Department of Mathematics,
Aringar Anna Government
Arts College for Women, Walajapet
email: dgthomasmcc@yahoo.com

Formal Language Aspects of Natural Computing– 2006, pp. 209–214

A Note on Molecular Computing algorithms for finding the Shortest Path with Forbidden Pairs

R. Ramkumar, P. Shrivatsav and S. Krishna Kumar

1 Introduction

We propose algorithms using DNA and Peptide based Molecular Computing models for the "Shortest Path with Forbidden Pairs" problem, known to be NP-complete. The problem involves finding the shortest path between two vertices of an undirected graph, under the restriction that given a collection of pairs of vertices, the path should contain at most one of every pair of vertices in the collection. Familiarity with DNA and peptide computing is assumed.

2 Problem Statement

Given an undirected graph $G = (V, E)$, a natural number m, a collection of pairs of vertices $C = \{(a_1, b_1), \ldots, (a_m, b_m)\}$ where $a_i, b_i \in V\ \forall i \in 1 \ldots m$ and two vertices $s, f \in V$; find the shortest path from s to f in G that contains at most one vertex from each pair of vertices in C.

3 Solution using DNA strands

3.1 Requirements

To run the algorithm, a few molecules need to be synthesized initially. First, a unique 20-mer (A strand with 20 nucleotides) is needed for each of the vertices in the graph G. Let these strands be called $v_i\ \forall i \in V$. Then, for each vertex v in any of a_i or b_i in C, a strand $\overline{v}$, complimentary to the strand chosen for v initially, is required. Further, for each edge e_{ij} between the vertices v_i and v_j, the following should be constructed: Let the 20-mer for any vertex v be written as $v_1 v_2 \ldots v_{20}$. Let $\overline{n}$ represent the complement of any nucleotide n. Construct the 20-mer $e_{ij} = \overline{v_{j10}} \ldots \overline{v_{j1}}\,\overline{v_{i20}} \ldots \overline{v_{i11}}$. The reader might notice that this construction is similar to that used by Adleman in [1]. Since G is undirected, for each edge, in addition to Adleman's construction, the 20-mer $e_{ji} = \overline{v_{i10}} \ldots \overline{v_{i1}}\,\overline{v_{j20}} \ldots \overline{v_{j11}}$ is needed.

<table>
<tr><td>e_{12}</td><td>GATCCAAATGTAATCCATAC</td></tr>
<tr><td></td><td>||||||||||||||||||||</td></tr>
<tr><td>v_1</td><td>ATGAATCTAGGTATGGATTA|||||||||| ATGAATCTAGGTATGGATTA</td></tr>
<tr><td>v_2</td><td>CATTTGGATCCGGGATACTA||||||||||</td></tr>
<tr><td></td><td>||||||||||||||||||||</td></tr>
<tr><td>e_{21}</td><td>TAATCCATACGATCCAAATG</td></tr>
</table>

Table 1: Example for generating edge strands

3.2 The Algorithm

Please refer Algorithm 1 for a concise description of the algorithm. We shall elaborate each of the steps in the following paragraphs:

Algorithm: ShortForbidDNA

Input: An undirected graph $G = (V, E)$, a natural number m, a collection of pairs of vertices $C = \{(a_1, b_1), \ldots, (a_m, b_m)\}$ where $a_i, b_i \in V\ \forall i \in 1 \ldots m$ and two vertices $s, f \in V$

Output: The shortest path, satisfying the constraints given, between s and f, and error in case there isn't any.

`// generate all possible paths`

mix all the v_i and e_{ij} to *synthesize* all paths possible

amplify paths generated using primers $\overline{s}$ and $\overline{f}$

select strands of *length* less than $n = |V|$

$S \longleftarrow$ corresponding single strands

`// check for forbidden pairs`

foreach $k \in 1 \ldots m$ **do**

  $S_1 \longleftarrow -(S, a_k)$

  $S \longleftarrow +(S, a_k)$

  $S_2 \longleftarrow -(S, b_k)$

  $S \longleftarrow S_1 \cup S_2$

end

`// find the smallest of the lot`

$S \longleftarrow$ *smallest*(S)

detect(S)

Algorithm 1: Shortest Path with Forbidden Pairs: DNA algorithm

Generate all paths

The first step of the algorithm involves generation of all possible paths in the graph. This we achieve by combining all v_i and e_{ij} in a test tube. For every edge between vertices i and j, this allows the strand $v_i v_j$ to hybridize with e_{ij}, leaving sticky ends of ten nucleotides each of either end. This arrangement allows for strands to be generated for every path in the graph. Consider, for example, a path from vertex 1 to 4 to 7. This allows the strand $v_1 v_4 v_7$ hybridized with $e_{14} e_{47}$ to form.

```
                      v1                      v4                      v7
5' <-- ATGAATCTAG GTATGGATTA CATTTGGATC CGGGATACTA TTAAGCTAGC CTAGGGTTAC --> 3'
                  |||||||||| |||||||||| |||||||||| ||||||||||
            3' <-- CATACCTAAT GTAAACCTAG GCCCTATGAT AATTCGATCG --> 5'
                          e14                     e47
```

Since the hybridization reaction is instantaneous, it can safely assumed that all valid paths in the graph are generated in this process.

3.2.1 Prune the sample space

The previous step also generates many paths which cannot be possible solutions for any instance of the problem at hand. Particularly, paths not starting from v_i and not ending with v_f violate the premise. Further, any path of length greater than $|V|$ cannot be a solution, as by pigeon-hole principle, some edge of the path has to be repeated and this indicates a loop. Any path with a loop cannot be the shortest path, as the path got by removing the edges in the loop satisfies the conditions in the problem statement as well, and is surely shorter than the path under consideration.

The former condition is enforced by doing a PCR first with $\overline{v_i}$, and then with $\overline{v_f}$ as the primers. This ensures that the strands amplified begin with v_i and end with v_f. PCR is capable of amplifications of an order of million times, and hence, the number of unamplified strands (the non-solutions) can be considered negligible. It is advisable, as is considered standard practice in molecular computing algorithms, to counter-check the results for correctness, and perhaps run the algorithm multiple times if is imperative that the optimal solution be obtained.

The second pruning step is achieved by Gel Electrophoresis. This procedure is known to biotechnologists as the *Agarose Gel* technique. In this, the strands are placed in a gel and subject to an electric potential. As a result, the strands "migrate" across the gel. It is known that the extent of migration is inversely proportional to the molecular weight of the species migrating. It is obvious that strands with greater length have a greater molecular weight as well. Hence, strands of length less than $|V|$ can be separated by selecting those which migrate by an extent greater than a minimum experimentally determinable cutoff.

It is to be noted that we get hybridized strands as a result of these steps. It is required that these strands be "melted" (heated to a sufficiently high temperature) to get back single strands, needed for the remaining steps of the algorithm.

3.2.2 Preventing the forbidden pairs

This is the bottleneck of the algorithm, a loop which checks for the presence of both vertices of some element in C, and removes them from the list of probables which could possibly be a solution to the problem. The experimental technique used for this is the magnetic bead separation technique. For every sub-sequence w to be tested for presence in the given strand, magnetic beads with $\overline{w}$ are embedded in a column and the strands allowed to pass through it. Strands with w in them hybridize with the $\overline{w}$ in the magnetic bead and stick to the column, while the rest flow down the column. Thus, those which emerge from the column do not contain w, while the strands stuck to the bead could be removed and molten to undo any hybridization; to get the strands containing w.

For each element $(a_i, b_i) \in C$, we pass the strands through magnetic beads containing a_i, and then b_i. Those which get stuck in both cases indicate the strands to be removed from the solution set, as per the problem specification.

3.2.3 Selecting the shortest path

All strands left behind after all the previous steps satisfy the forbidden pair conditions and are paths from s to f. The last step of the algorithm involves selecting the shortest strands of them all, as we need the shortest path. As expected, it is Gel Electrophoresis, as described in Section 3.2.1, which comes to our rescue. The strands which migrate the maximum need to be selected, as they have the smallest molecular weight.‡

3.3 Complexity

The bottleneck of the algorithm lies in the loop checking for the absence of both nucleotides for every pair of vertices in the given collection. The remaining steps, viz. path generation, sample space pruning and shortest path selection take constant time only. Thus, the algorithm takes $O(m)$ time.

4 Solution using peptides

4.1 Requirements

We first need a few peptides and antibodies as the "apparatus" for the algorithm. First, for each vertex of G, say v_i, we need an unique epitope. For the sake of simplicity, we shall henceforth refer to both the vertex and its epitope by the same name — the context should make the meaning clear. Further, for every pair of vertices i, j, an antibody targeting the epitope $v_i v_j$ is needed. Let us call this antibody A_{ij}, if v_i and v_j are adjacent, and B_{ij} otherwise. It is required that affinity(B_{ij}) > affinity(A_{kl}), $\forall i, j, k, l$. Next, antibodies (D_k, E_k) are needed against the epitopes corresponding to the vertices (a_k, b_k), $\forall\ (a_k, b_k) \in C$. The antibodies chosen are such that affinity(D_k), affinity(E_k) > affinity(A_{kj}) $\forall$ antibodies A_{kj}, $\forall$ $k \in 1 \dots m$.

4.2 The Algorithm

The algorithm is similar in many respects to the DNA algorithm discussed earlier. Algorithm 2 gives a compact outline of the entire process. The next few paragraphs describe the processes involved in greater detail.

Generate all paths

First, we need to generate all possible paths in the graph. For this purpose, we generate peptide sequences representing a permutation of an arbitrary subset of the vertices in the graph, such that the initial and the final vertices form the ends of the sequence. Further, except for the initial and the final vertex, all other vertices in the sequence are represented by duplicated epitopes. Each of these sequences depict a possible path between the initial and the final vertex. For example, for a five vertex graph, a sequence $v_1 v_2 v_2 v_4 v_4 v_5$ represents the possible path $e_{12} e_{24} e_{45}$. Note that all vertices need not be present in a given sequence, and in fact, more than one pair of the same vertex might occur. Further, some of these edges might not be present in the graph under consideration. In our example, e_{24} might not exist in the graph.

‡Experimentally, we need to take care of the fact that the nucleotides can be of different molecular weights, and hence the weights could be slightly different for the same length. Thus, we might have to select strands from an, albeit narrow, range of migration distances. However, since we use 20-mers as the fundamental unit, we can be sure that in this process, we don't select strands of a greater length than the shortest.

```
Algorithm:ShortForbidPep
Input: An undirected graph G = (V, E), a natural number m, a collection of pairs of
       vertices C = {(a_1, b_1), . . . , (a_m, b_m)} where a_i, b_i ∈ V ∀i ∈ 1 . . . m and two
       vertices s, f ∈ V
Output: The shortest path, satisfying the constraints given, between s and f, and
        error in case there isn't any.
// generate all possible paths
generate all possible peptides starting with s, ending with f, and having doubly
duplicated permutations of all the vertex epitopes
add all antibodies A_ij
attach all B_ij to magnetic beads
pass through column, call the strands coming out unattached, S
// check for forbidden pairs
foreach k ∈ 1 . . . m do
    pass S through column lined with D_k
    S_1 ⟵ strands which run off
    elute column
    pass through column lined with E_k
    S_2 ⟵ strands which run off
    S ⟵ S_1 ∪ S_2
end
// find the smallest of the lot
S ⟵ smallest(S)
detect(S)
```

Algorithm 2: Shortest Path with Forbidden Pairs: Peptide algorithm

Our next job therefore, is to weed out those peptides which have non-existent edges (The sequences with repeated vertices are automatically weeded out when the shortest path is selected). This process is done by separation using a column and magnetic beads, and is conceptually similar to the technique used for DNA strands. Here, the process is called *elution*. The peptides passing out unbound are called the *eluents* and are said to have *eluted* out. First, we add to the peptides, all antibodies A_{ij}. Since there exists such an antibody for each pair of adjacent vertices, these will bind to those parts of the peptides which represent genuine edges. Those which have fake edges will have some portions of them not bound to one of the A_{ij}. Next, we pass the bound strands through a column packed with magnetic beads containing all B_{ij}. There is a B_{ij} antibody for each fake edge, and hence antibodies with fake edges will get bound to the column, while genuine paths get eluted out. Thus, the peptides which come out all represent genuine paths in G, from s to f.

Preventing the forbidden pairs

This step is almost similar to the approach used in the DNA algorithm. Here too, a column is used, and for each pair of vertices (a_k, b_k) in C, the corresponding antibodies (D_k, E_k) bind to the peptides to indicate the presence of the two vertices. First, the peptides are eluted against D_k, and the bound peptides are further eluted against E_k. The eluents at any stage indicate the peptides which do not have at least one of the two vertices, and hence, can become a solution for the problem. Such peptides can carried over for testing against the next pair in the forbidden collection.

Selecting the shortest path

All the peptides from the previous step satisfy the forbidden pairs condition. The last step therefore, is to select the shortest strand of them all. A Gel Electrophoresis for the peptides (a technique known as *SDS PAGE gel*) leaves the solutions to the problem farthest off from the starting point of migration (maximum migration).

As was the case with the DNA algorithm, the peptide algorithm takes $O(m)$ time as well, the linearity being due to the single loop preventing any of the forbidden pairs from being present in the solution. The other steps take constant time, and hence do not count towards the complexity measure.

5 Conclusion

Most molecular computing algorithms, including the ones discussed above, are primarily of theoretical interest. There are many practical issues, which need to be looked at, and hopefully solved, before these algorithms become experimentally feasible.

References

[1] Leonard M. Adleman. Molecular computation of solutions to combinatorial problems. *Science*, 266(11):1021–1024, 1994.

R. Ramkumar
Department of Computer Science and Engineering
Indian Institute of Technology Madras
Chennai - 600 036, India

P. Shrivatsav and S. Krishna Kumar
Department of Biotechnology
Indian Institute of Technology Madras
Chennai - 600 036, India
email: andyetitmoves@gmail.com

Formal Language Aspects of Natural Computing– 2006, pp. 215–219

A Note on Peptide Computing Solution for the `NP-Complete` Edge Coloring Problem

R. Rahul and Suneeth Reddy

1 Introduction

In Graph Theory, an edge coloring when mentioned without any qualification is always assumed to be a **proper** edge coloring of the graph, meaning no two **adjacent** edges are assigned the same color. A proper edge coloring with k colors is called a (proper) ***k*-edge-coloring** and is equivalent to the problem of partitioning the edge set into k matchings. The smallest number of colors needed in a (proper) edge coloring of a graph G is the **chromatic index**, or *edge chromatic number*, $\chi'(G)$, also sometimes notated $\chi_1(G)$.
A few properties of $\chi'(G)$:

1. $\chi'(G) \geq \Delta(G)$.

2. $\chi'(G) \leq \Delta(G) + 1$. (Vizing's theorem [1])

Here $\Delta(G)$ is the maximum degree;

As we can see from above, $\chi'(G)$ equals either $\Delta(G)$ or $\Delta(G) + 1$. When $\chi'(G) = \Delta(G)$, G is said to be **Class 1**; otherwise, **Class 2**. Holyer (1981) [2] proved that it is `NP-Complete` to determine whether a simple graph is Class 1 or Class 2.

2 Reduction of this problem to other `NP-complete` problems

The edge coloring problem can be reduced to the vertex-coloring problem, by transforming the graph G to another graph G', where the edge set $\{e_i\}$ is transformed to the vertex set $\{v_i'\}$ and if (e_i, e_j) are adjacent in graph G, then e_{ij}' is the edge joining the vertices v_i' and v_j' in graph G'.

3 The solution to this problem via peptide computing

Here we give a solution to this `NP-Complete` problem without reducing it to any of the other `NP-Complete` problem. Since $\chi'(G)$ equals either $\Delta(G)$ or $\Delta(G)+1$, our method takes $\Delta(G)$ colors as an input value to test if the graph can be colored so. Else, $\chi'(G) = \Delta(G) + 1$.

Let the graph G have n vertices V_1 to V_n, and m edges e_1 to e_m. Let the Δ colors be $C_1, \ldots, C_\Delta$.

4 The process

Without loss of generality, we can assume that the vertex V_1 has the degree Δ. Also, let us assume that the edges from this vertex are $e_1, \ldots, e_\Delta$. The remaining edges in the graph thus are $e_{\Delta+1}, \ldots, e_m$.

The idea that we will be implementing is represented below.

We will assume that $e_1, \ldots, e_\Delta$ are colored with $C_1, \ldots, C_\Delta$ respectively. Then we will check if it is possible to group the remaining edges under each of these Δ edges.

Now the problem lies in finding out if such a grouping exits.

C_1	C_2	....	C_Δ
e_1	e_2	....	e_Δ
$e_j, e_k, \ldots$	$e_l, e_m, \ldots$	....	$e_x, e_y, \ldots$..

5 Epitopes

- Each edge e_i has a corresponding epitope E_i.
- There is a general epitope E, which is different from all other epitopes.
- There is a terminal epitope E_F.
- Peptides will be so designed so that they can be thought to be made of mini peptides of length Δ.

The peptide can be envisaged in the following way:

$$E_{a1}E_{a2}\ldots E_{a\Delta}\vdots E_{b1}E_{b2}\ldots E_{b\Delta}\vdots\ldots\ldots\ldots\vdots E_{c1}E_{c2}\ldots E_{c\Delta}$$

- Each of these blocks above is of length Δ
- Each epitope above should belong to the set $\{E_1, \ldots, E_m\}$ U $\{E\}$.
- Many copies of E can exist within the peptide (repeating epitope). Only one copy of each of the $E_i(i \ \varepsilon \ \{1, \ldots, m\})$ can exist in each peptide. That is, $E_1, \ldots, E_m$ are permuted among the peptides

The idea is that E_{a1}, E_{b1} and so on have the first color. E_{a2}, E_{b2}, and so on have the second color. Since each of these envisaged blocks has exactly Δ elements, we are trying to group the edges into Δ groups, with each group being assigned one color.

Each peptide would start with the sequence $E_1E_2E_3\ldots E_\Delta$, so essentially, the first envisaged block is made of the Δ edges adjacent to the vertex of degree Δ. The remaining blocks can comprise of the remaining E_i's or E's. However atleast one E_i should be present in every block.

Thus the maximum length of the peptide would be:

$$\Delta + (m - \Delta)\Delta$$

This can be realized by designing the initial solution in such a way, that all peptides start with the leader sequence $E_1E_2E_3\ldots E_\Delta$, and the remaining peptide is a permutation over the set { $[E_{\Delta+1}], \ldots, [E_m]$ }, where each $[E_i]$ denotes one epitope from the set:

$$\{E_i, EE_i, EEE_i, \ldots\ldots, E^{\Delta-1}E_i\}.$$

6 Design of Antibodies

The antibodies are designed with the intention of identifying those edges (epitopes) at positions i, $(i + j\Delta)$, that share a common vertex. This can be achieved by designing antibodies that bind to minipeptides of lengths $\Delta + 1, 2\Delta + 1, 3\Delta + 1, \ldots, (m - \Delta)\Delta + 1$, and have specificities only at the ends. In other words, these antibodies identify a pair of eptiopes only if they have affinity towards both of them (share a common vertex) and have coordinates i, $(i + j\Delta)$ (are colored with the same color).

Formally

For every vertex V_i of degree δ_i, there should exist $(m - \Delta) *^{\delta_i} P_2$ antibodies which will have affinity towards the edges adjacent to V_i.

Thus A_{ijkl} ($i\ \varepsilon\ \{1, \ldots, n\}$ and $(j, k)\ \varepsilon\ \{1, \ldots, \delta_i\}$ and $l\ \varepsilon\ \{1, \ldots, m - \Delta\}$) would recognize mini-peptides of the form:

$$E_{ij}\ \alpha(\{E_i, i\varepsilon\{1, \ldots, m\}\}\ U\ \{E\}^*)E_{ik}$$

α(M) denotes permutations from the elements of M.

and $|\alpha(M)| = (l\Delta - 1)$ and e_{ij}, e_{ik} should be adjacent to V_i.

Lastly, every peptide would end with a common antibody E_F.

Among all these peptides formed, the minimum length would be m + 1 (= I (say)) (no copy of E at all within the peptide).

The maximum length on the other hand would be $\Delta + (m - \Delta)\Delta + 1$ (= F (say)). This would be the case when every envisaged block has only one edge (apart from the initial leading sequence).

7 Fluorescent Antibodies

Another set of fluorescent antibodies $B_i \varepsilon\{B_I, \ldots, B_F)$ are to be designed such that,

- B_i would identify all epitopes of the form:

$$E_1\ \alpha(\{E_i, i\ \varepsilon\ \{1, \ldots, m\}\}\ U\ \{E\}^*)\ E_F$$

$|\alpha(M)| = i - 2$.

- B_i would have lower binding affinity than any A_i.
- Thus, to a peptide, even if one A_i binds, no B_i can bind to that peptide.

8 The Operation

- Form all peptides as specified earlier (Initial Solution Space).
- Add antibodies A_{ijkl}

$$i\varepsilon\{1, \ldots, n\} \text{ and } (j, k)\varepsilon\{1, \ldots, \delta i\} \text{ and } l\varepsilon\{1, \ldots, m - \Delta\}$$

- Add fluorescent antibodies $B_i \varepsilon \{B_1, \ldots, B_F)$, which are of lower affinity than A_{ijkl}
- If fluorescence is detected then Δ coloring exists.
- If peptides are bound to an addressed chip the solution can be immediately read.

9 Complexity

This model requires:

- Number of peptides $= \Delta(m - \Delta)(m - \Delta)!$
- Max Length of peptides $= \Delta + (m - \Delta)\Delta + 1 = O(m\Delta)$
- Number of antibodies

$$\leq n(m - \Delta)\Delta(\Delta - 1) + (m - \Delta + 1)\Delta + 1 - m = O(nm\Delta^2)$$

10 Example

Consider the graph shown below:

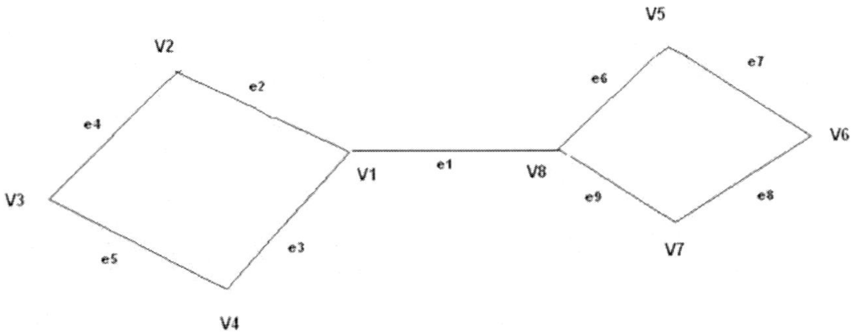

The graph has 8 vertices (n = 8), 9 edges (m = 9) and $\Delta = 3$

- Peptide 1 (Assignment 1)

  $E_1E_2E_3 \vdots E_4E_5\ \underbrace{E_6 \vdots E_7E_8E_9}\ \vdots E_F$

  Antibody A_{8691} would bind, hence not a correct assignment.

- Assignment 2

  $\underbrace{E_1E_2E_3 \vdots EEE_4 \vdots E_9}\ E_8E_7 \vdots E_6EE \vdots E_5E_F$

  Antibody A_{8192} would bind to this one, hence this is not a correct assignment.

- Assignment 3

  $E_1E_2E_3 \vdots E_5EE \vdots E_8E_7E_4 \vdots EEE_6 \vdots EE_9E_F$

  No A Antibody will bind to it. Hence the fluorescence antibody, B_{15} (in this case) will bind to it.

References

[1] S. Fiorini and R.J. Wilson: Edge-colourings of Graphs, Pitman, London, 1977.

[2] Ion Holyer: The NP-Completeness of Edge-Colouring. *Siam J. Comput* **10(4)** Nov. 1981, 718–720.

[3] M. Sakthi Balan, K. Krithivasan and Y. Sivasubramaniam: Peptide Computing - Universality and Complexity. *Lecture Notes in Computer Science* **2340** Springer-Verlag, 2001, 290–299.

R. Rahul
Department of Biotechnology
Indian Institute of Technology Madras
Chennai 600 036, India

Suneeth Reddy
Department of Computer Science and Engineering
Indian Institute of Technology Madras
Chennai 600 036, India

Formal Language Aspects of Natural Computing– 2006, pp. 221–224

A note on DNA and Peptide Computing: Knapsack Problem

B. Naveen Kumar, Anshu Bhatia and Sourav Chatterjee

Statement of Problem

"Given a set of items, each with a cost and a value, can a value of at least V be achieved without exceeding the cost C?"

Approach:

The nature of the problem suggests that the cost and value be represented by two independent properties of DNA / peptide strands which add up on concatenation of the strands. The summing up of one property should not have any bearing upon the other property.

DNA Computing

Representation:

Cost: The length of a single DNA strand is used to represent cost. Thus greater the length of a strand greater is the cost it is representing.

Value: A binary representation is used for value. Thus the rightmost bit is represented by a specific sequence of two nucleotides. Any sequence other then that particular two nucleotide sequence represents a 0 at the rightmost bit. The next 1 is represented by a specific sequence of four nucleotides. Again anything other then the particular four nucleotide sequence represents a 0 at the bit next to the rightmost bit. The 1 for the next bit two places left to the rightmost one is represented by a specific eight nucleotide sequence.

Thus the number of nucleotides representing 1s is two more then the sum of the number of all other nucleotides representing 1s to the right of it. Thus the rightmost 1 is represented by a sequence of 2 nucleotides, the next 1 by a sequence of 4 nucleotides, the 1 beyond by a sequence of 8 nucleotides and so on.

In this way the item representing the maximum value has a specific binary representation which corresponds to a maximum length of single strand DNA for representing value. All other values are represented by strands of the same length as the strand representing the maximum value.

Thus in effect, each item represented by a single stranded DNA with 2 parts - a constant length part (value) and a variable length part (cost).

As an example consider the cost and value representation as shown here.

Item 1 : AGTCA : ACTAGGCTXXXXGT
Item 2 : AGTCAGTC : XXXXXXXXGATCGT
Item 3 : AGTCAG : ACTAGGCTGATCXX

ITEM	COST	VALUE
1	5	5
2	8	3
3	6	6

Table 1: Representation of cost and value of 3 items

DNA strand representation of the items

The portion of the strand to the left of the : (colon) represents the cost. GT represents 1 at the rightmost bit, GATC represents 1 at the 2nd bit, and ACTAGGCT represents 1 at the 3rd bit. Xs represent 0s and are sequences other then the sequences representing 1s. Note that the length of the strand to the right of the: is same in all the 3 items. Thus the representation ensures that the 2 properties representing cost and value are independent of each other.

DNA Algorithm:

1. All single stranded DNA oligonucleotides are mixed together.
2. All possible combinations of the single strands are formed.
3. Separation on the basis of total cost (length).
4. Combinations below the threshold cost (C) selected.
5. Check whether any of these combinations has a value greater than or equal to the threshold value.

Step 2 can be performed with the help of a ligase enzyme, which will ligate the single stranded DNAs in all possible fashions.

Step 3 and 4 have to be performed by agarose gel electrophoresis, which separates strands based upon their molecular weights. Molecular weight of strands (lengths) represents cost over here.

Step 5 has to be performed by passing all the strands from step 4 through a column containing sequences complementary to the value sequence bits. A strand representing the maximum value will be eluted in the end through such a column.

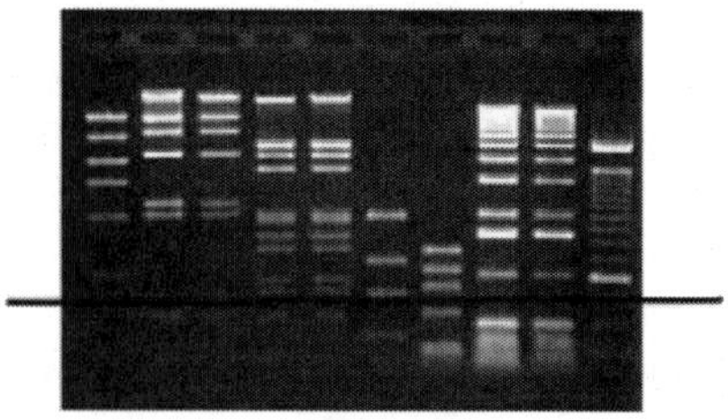

Figure 1: A garose gel electrophoreesis: Strands below the line represent cost less than a particular threshold cost

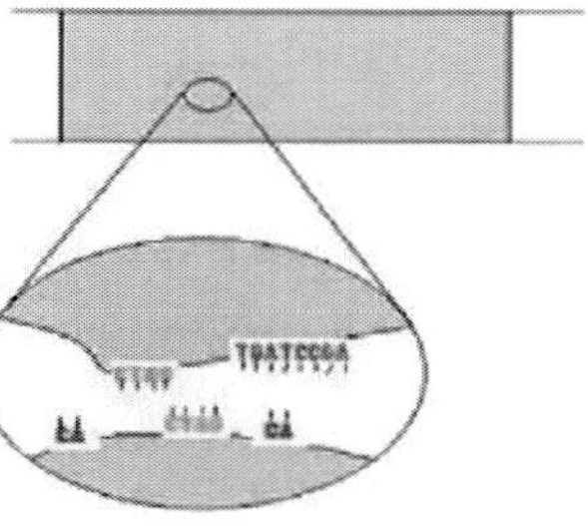

Figure 2: Column containing sequences complementary to the sequences representing 1s at different bit places. The strand representing the highest value binds to the column with greatest affinity and is the last one to get eluted. This column helps in checking whether or not a value greater than a particular value exists.

Peptide computing:

Amino acids have side chains which can be polar or non-polar, charged or uncharged and hydrophobic or hydrophilic [Biochemistry, Stryer's 4th edition]. So in addition to mass of peptides, any of the above properties can also be used for the purpose of representation in this case. Since, existing technology allows us to quantify only the charge on peptide chains, we have used length/mass and charge of peptide chains as the two properties representing cost and value respectively. Thus similar to DNA computing,

Representation:

Cost : represented by no. of non polar amino acids (varying part of each peptide)

Value : represented by no. of acidic amino acids (fixed length part of each peptide)

Thus, each item represented by a peptide strand with 2 parts - a constant length part (value) and a variable length part (cost). As an example consider the following 3 items.

ITEM	COST	VALUE
1	5	5
2	8	3
3	6	6

Item 1 : MPWVV : LPDWEGADDE
Item 2 : GALIMFMA : PGAMDEEMWL
Item 3 : LIAVMF : GEEIDMLDDD

Each of the symbols is the one lettered code for one of the 20 amino acids [Biochemistry, Stryer's 4th edition. The portion the strand to the left of the: represents the cost and is varying in length. The portion of the strand to the right of the: represents the value. Both D and E are negatively charged amino acids and presence of one D indicates a value of 1, presence of 2 Ds indicates a value of 2 and so on. Again, note that the length of the strand to the right of the: is same in all the 3 items. Thus the representation ensures that the 2 properties representing cost and value are independent of each other.

Peptide Algorithm:

1. All peptides are mixed together
2. All possible combinations are formed
3. Separation on the basis of total cost (length)
4. Combinations below the threshold cost (C) selected.
5. Check whether any of these combinations has a value (charge) greater than or equal to the threshold value.

Again step 3 and step 4 can be performed with the help of gel electrophoresis. In the case of peptides, polyacrylamide gel is used for separating strands based upon their molecular weights, which represent cost over here.

Step 5 can be performed with the help of chromatographic columns which separate strands based upon their charge. Thus such a column will help us ascertain whether or not a value strand exceeding the decided value exists or not.

It should be noted that only suggestions are given here for the implementation of the algorithm.

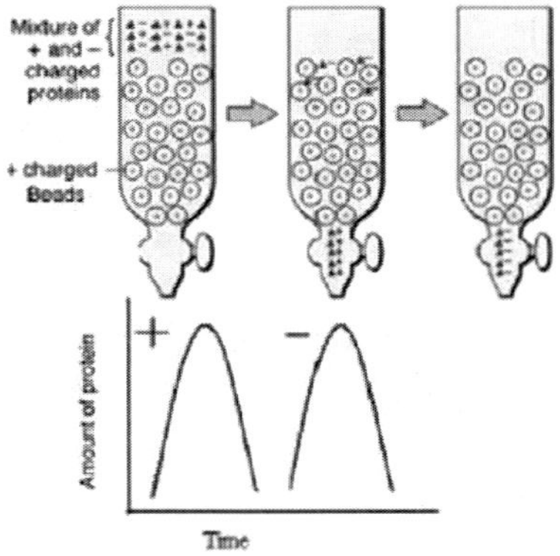

Figure 3: A chromatographic column containing positively charged beads. Note that positively charged peptides get eluted first and then negatively charged peptides. Thus the peptides representing the greatest value are eluted at the end. This column helps in checking whether or not a value greater than a particular exists.

B. Naveen Kumar and Anshu Bhatia
Department of Biotechnology
Indian Institute of Technology Madras
Chennai - 600 036, India

Sourav Chatterjee
Department of Mechanical Engineering
Indian Institute of Technology Madras
Chennai - 600 036, India

Author Index

Made in United States
North Haven, CT
05 November 2026

21772753R00128